# FORCEnet
## IMPLEMENTATION STRATEGY

Committee on the FORCEnet Implementation Strategy
Naval Studies Board
Division on Engineering and Physical Sciences

NATIONAL RESEARCH COUNCIL
*OF THE NATIONAL ACADEMIES*

THE NATIONAL ACADEMIES PRESS
Washington, D.C.
**www.nap.edu**

**THE NATIONAL ACADEMIES PRESS   500 Fifth Street, N.W.   Washington, DC 20001**

NOTICE: The project that is the subject of this report was approved by the Governing Board of the National Research Council, whose members are drawn from the councils of the National Academy of Sciences, the National Academy of Engineering, and the Institute of Medicine. The members of the committee responsible for the report were chosen for their special competences and with regard for appropriate balance.

This study was supported by Contract No. N00014-00-G-0230, DO #17, between the National Academy of Sciences and the Department of the Navy. Any opinions, findings, conclusions, or recommendations expressed in this publication are those of the author(s) and do not necessarily reflect the views of the organizations or agencies that provided support for the project.

International Standard Book Number 0-309-10025-9 (Book)
International Standard Book Number 0-309-68385-1 (PDF)

Copies of this report are available from:

Naval Studies Board
The Keck Center of the National Academies
500 Fifth Street, N.W., Room WS904
Washington, DC 20001

Additional copies of this report are available from the National Academies Press, 500 Fifth Street, N.W., Lockbox 285, Washington, DC 20055; (800) 624-6242 or (202) 334-3313 (in the Washington metropolitan area); Internet, http://www.nap.edu.

# THE NATIONAL ACADEMIES
*Advisers to the Nation on Science, Engineering, and Medicine*

The **National Academy of Sciences** is a private, nonprofit, self-perpetuating society of distinguished scholars engaged in scientific and engineering research, dedicated to the furtherance of science and technology and to their use for the general welfare. Upon the authority of the charter granted to it by the Congress in 1863, the Academy has a mandate that requires it to advise the federal government on scientific and technical matters. Dr. Ralph J. Cicerone is president of the National Academy of Sciences.

The **National Academy of Engineering** was established in 1964, under the charter of the National Academy of Sciences, as a parallel organization of outstanding engineers. It is autonomous in its administration and in the selection of its members, sharing with the National Academy of Sciences the responsibility for advising the federal government. The National Academy of Engineering also sponsors engineering programs aimed at meeting national needs, encourages education and research, and recognizes the superior achievements of engineers. Dr. Wm. A. Wulf is president of the National Academy of Engineering.

The **Institute of Medicine** was established in 1970 by the National Academy of Sciences to secure the services of eminent members of appropriate professions in the examination of policy matters pertaining to the health of the public. The Institute acts under the responsibility given to the National Academy of Sciences by its congressional charter to be an adviser to the federal government and, upon its own initiative, to identify issues of medical care, research, and education. Dr. Harvey V. Fineberg is president of the Institute of Medicine.

The **National Research Council** was organized by the National Academy of Sciences in 1916 to associate the broad community of science and technology with the Academy's purposes of furthering knowledge and advising the federal government. Functioning in accordance with general policies determined by the Academy, the Council has become the principal operating agency of both the National Academy of Sciences and the National Academy of Engineering in providing services to the government, the public, and the scientific and engineering communities. The Council is administered jointly by both Academies and the Institute of Medicine. Dr. Ralph J. Cicerone and Dr. Wm. A. Wulf are chair and vice chair, respectively, of the National Research Council.

**www.national-academies.org**

## COMMITTEE ON THE FORCEnet IMPLEMENTATION STRATEGY

RICHARD J. IVANETICH, Institute for Defense Analyses, *Co-Chair*
BRUCE WALD, Arlington, Virginia, *Co-Chair*
ROBERT F. BRAMMER, Northrop Grumman Information Technology
JOESPH R. CIPRIANO, Lockheed Martin Information Technology
ARCHIE R. CLEMINS, Caribou Technologies, Inc.
BRIG "CHIP" ELLIOTT, BBN Technologies
JOEL S. ENGEL, Armonk, New York
JUDE E. FRANKLIN, Raytheon Network-Centric Systems
JOHN T. HANLEY, JR., Institute for Defense Analyses
KERRIE L. HOLLEY, IBM Global Services
KENNETH L. JORDAN, JR., Cabin John, Maryland
OTTO KESSLER, The MITRE Corporation
JERRY A. KRILL, Applied Physics Laboratory, Johns Hopkins University
ANN K. MILLER, University of Missouri-Rolla
WILLIAM R. MORRIS, Alexandria, Virginia
RICHARD J. NIBE, Amelia Island, Florida
JOHN E. RHODES, Balboa, California
DANIEL P. SIEWIOREK, Carnegie Mellon University
EDWARD A. SMITH, JR., The Boeing Company
MICHAEL J. ZYDA, University of Southern California

### Staff

CHARLES F. DRAPER, Director, Naval Studies Board
MICHAEL L. WILSON, Study Director (through August 27, 2004)
SUSAN G. CAMPBELL, Administrative Coordinator
MARY G. GORDON, Information Officer
IAN M. CAMERON, Research Associate
AYANNA N. VEST, Senior Program Assistant
SIDNEY G. REED, JR., Consultant

**Staff**

CHARLES F. DRAPER, Director
ARUL MOZHI, Senior Program Officer
SUSAN G. CAMPBELL, Administrative Coordinator
MARY G. GORDON, Information Officer
IAN M. CAMERON, Research Associate
AYANNA N. VEST, Senior Program Assistant

# Preface

Visionary Navy leaders enunciated the tenets of network-centric operations beginning in the early 1990s, and in 1998 requested the advice of the Naval Studies Board of the National Research Council (NRC) about how to achieve such capabilities. The resulting report was entitled *Network-Centric Naval Forces: A Transition Strategy for Enhancing Operational Capabilities.*[1] Although the Navy adopted some of the recommendations from that report—notably the establishment of what became the Naval Network Warfare Command—progress was limited on many fronts until the Chief of Naval Operations (CNO) Strategic Studies Group described a networked, distributed, combat force as a "FORCE-net."[2] The CNO incorporated the FORCEnet concept into Sea Power 21[3]—the overall vision for transforming the Navy—and adopted the following definition of FORCEnet:

> [FORCEnet is] the operational construct and architectural framework for naval warfare in the information age that integrates warriors, sensors, networks, com-

---

[1]The report defined network-centric operations as "military operations that exploit state-of-the-art information and networking technology to integrate widely dispersed human decision makers, situational and targeting sensors, and forces and weapons into a highly adaptive, comprehensive system to achieve unprecedented mission effectiveness." Naval Studies Board, National Research Council. 2000. *Network-Centric Naval Forces: A Transition Strategy for Enhancing Operational Capabilities*, National Academy Press, Washington, D.C., p. 1.

[2]ADM James R. Hogg, USN (Ret.), Director, CNO Strategic Studies Group, personal communication, November 9, 2005.

[3]ADM Vern Clark, USN. 2002. "Sea Power 21 Series, Part I: Projecting Decisive Joint Capabilities," *U.S. Naval Institute Proceedings*, October.

mand and control, platforms, and weapons into a networked, distributed, combat force that is scalable across all levels of conflict from seabed to space and sea to land.[4]

Although this definition views FORCEnet as the operational construct and the architectural framework for the entire transformed Navy, some have viewed FORCEnet merely as an information network and the associated FORCEnet architecture merely as an information systems architecture. In the first view, the FORCEnet architecture would affect the functional allocation across all naval systems; in the latter view it would only impose a standard data interface on these systems. Furthermore, although FORCEnet is not a system, the Navy's requirements-formulation and materiel-acquisition organizations have tended to view FORCEnet as a set of individual information systems that can be developed and acquired by traditional methods.

To assist the Navy in better defining its approach to FORCEnet, the Department of the Navy asked the NRC's Naval Studies Board to conduct a study that would provide a recommended FORCEnet implementation strategy. The specific terms of reference for this study are presented in Chapter 8 along with cross-references to the committee's recommendations.

## THE COMMITTEE'S APPROACH

The approach of the Committee on the FORCEnet Implementation Strategy[5] was to organize itself around the specific operational, policy, and technical areas necessary to fulfill the tasks laid out in the terms of reference. The committee first convened in September 2003, holding additional meetings over a period of 7 months, both to gather input from the relevant communities and to discuss the committee's findings.[6] The agendas for the meetings from September 2003 through March 2004 are provided in Appendix B. The months between the last meeting and publication of the report were spent preparing the draft manuscript, gathering additional information, reviewing and responding to the external review comments, editing the report, and conducting the required security review necessary to produce an unclassified report.

---

[4]VADM Richard W. Mayo, USN; and VADM John Nathman, USN. 2003. "Sea Power 21 Series, Part V: FORCEnet: Turning Information into Power," *U.S. Naval Institute Proceedings*, February, p. 42.

[5]Brief biographies of all committee members are presented in Appendix A.

[6]During the course of its study, the committee held meetings at which it received (and discussed) classified materials. Accordingly, the content of the present report is limited because of restrictions that apply to the use of classified information.

## STRUCTURE AND CONTENT OF THIS REPORT

Chapter 1 of this report presents a scenario to illustrate a FORCEnet vision of fully networked operations, outlines the characteristics required for achieving this vision, discusses the status of network-centric capabilities, and warns of formidable challenges. The next five chapters address these challenges. Chapter 2 deals with the need for common understanding of the meaning of FORCEnet across the naval enterprise and urges acceptance of the CNO's definition. Chapter 3 describes the context of joint and Department of Defense plans and initiatives within which FORCEnet must be implemented, and recommends strong coupling of the concept to the combatant commanders. Chapter 4, in which it is accepted that FORCEnet has no fixed end state, deals with the challenges of implementing a complex system through the discussed coevolution of operational concepts and materiel. Chapter 5 deals with the challenge of engineering a complex system; notes the importance of controlling interfaces as capabilities evolve; embraces the network-centric checklist of the Assistant Secretary of Defense for Networks and Information Integration and the open architecture developed by the Naval Sea Systems Command and adopted by the Program Executive Officer for Integrated Warfare Systems; and urges the implementation of a distributed engineering plant for FORCEnet. Chapter 6 discusses potential capability shortfalls in the FORCEnet information infrastructure and recommends science and technology investments to overcome them. Chapter 7 collects the principal recommendations of the report, presenting them together with a concise version of the discussion and the findings that led to them. This chapter builds on the idea of an implementation strategy by incorporating the recommendations within a set of objectives required for such a strategy. Chapter 8 cross-references the committee's recommendations to the study terms of reference.

# Acknowledgment of Reviewers

This report has been reviewed in draft form by individuals chosen for their diverse perspectives and technical expertise, in accordance with procedures approved by the National Research Council's (NRC's) Report Review Committee. The purpose of this independent review is to provide candid and critical comments that will assist the institution in making its published report as sound as possible and to ensure that the report meets institutional standards for objectivity, evidence, and responsiveness to the study charge. The review comments and draft manuscript remain confidential to protect the integrity of the deliberative process. We wish to thank the following individuals for their review of this report:

Charles F. Bolden, Jr., MajGen, USMC (Ret.), Houston, Texas
Herbert A. Browne, VADM, USN (Ret.), Armed Forces Communications and Electronics Association,
Millard S. Firebaugh, RADM, USN (Ret.), General Dynamics, Electric Boat Corporation,
Charles M. Herzfeld, Silver Spring, Maryland,
Bill B. May, Los Altos Hills, California,
Cynthia R. Samuelson, LMI,
Fred B. Schneider, Cornell University, and
Michael G. Sovereign, Monterey, California.

Although the reviewers listed above provided many constructive comments and suggestions, they were not asked to endorse the conclusions or recommendations, nor did they see the final draft of the report before its release. The review of this report was overseen by Robert A. Frosch, Harvard University, and Robert J. Hermann, Global Technology Partners, LLC. Appointed by the NRC, they were responsible for making certain that an independent examination of this report was carried out in accordance with institutional procedures and that all review comments were carefully considered. Responsibility for the final content of this report rests entirely with the authoring committee and the institution.

# Contents

# Executive Summary

FORCEnet is the Department of the Navy's (DON's) approach for enhancing its capability to perform network-centric operations. The Chief of Naval Operations (CNO) has embraced the following definition of FORCEnet:

> [FORCEnet is] the operational construct and architectural framework for naval warfare in the information age that integrates warriors, sensors, networks, command and control, platforms, and weapons into a networked, distributed, combat force that is scalable across all levels of conflict from seabed to space and sea to land.[1]

The CNO has requested that the Naval Studies Board of the National Research Council (NRC) provide advice regarding both the adequacy of this definition and the actions required to implement FORCEnet (see Chapter 8 for the terms of reference). The Committee on the FORCEnet Implementation Strategy was formed by the NRC to respond to that request.

## APPROACH TO DEVELOPING RECOMMENDATIONS

The principal recommendations and a concise version of the arguments that support them are contained in Chapter 7 of this report.[2] That chapter builds on the idea of an implementation strategy by incorporating the committee's recommen-

---

[1]VADM Richard W. Mayo, USN; and VADM John Nathman, USN. 2003. "Sea Power 21 Series, Part V: ForceNet: Turning Information into Power," *U.S. Naval Institute Proceedings*, February, p. 42.

[2]Chapters 1 through 6 contain the full discussion of these issues.

dations within a set of objectives required for such a strategy. The committee emphasizes its belief that all of these recommendations are important and that the implementation of some of them should not preclude implementation of the others. This Executive Summary is a greatly condensed presentation of the committee's findings, organized under the eight implementation imperatives listed below. Accompanying the findings are extracts from the committee's detailed recommendations and footnotes referencing other detailed recommendations and supporting material.

## IMPLEMENTATION IMPERATIVES
## AND MAJOR RECOMMENDATIONS

The following implementation imperatives (addressed in individual subsections below) are distilled from a set of implementation strategy objectives presented in Chapter 7 of this report. The imperatives are necessary to establish a set of guiding principles for the Navy and Marine Corps to realize FORCEnet.

- Recognize that FORCEnet is more than an information network.
- Accept that FORCEnet has no fixed end state.
- Establish governance mechanisms that span the Office of the Chief of Naval Operations (OPNAV), the acquisition community, and the fleet.
- Devote more resources to developing operational constructs.
- Base resource allocation decisions on packages that reflect network-centric operational concepts.
- Strengthen architectural development and systems engineering capabilities.
- Strengthen the naval coupling to the combatant commanders.
- Exploit Global Information Grid (GIG) capabilities while preparing to fill GIG gaps and determining the limits of network-centricity.

### Recognize That FORCEnet Is More Than an Information Network

**Recommendation for OPNAV, Naval Network Warfare Command (NETWARCOM), and Marine Corps Combat Development Command (MCCDC): Articulate better the meaning of the terms "operational construct" and "architectural framework" in the description of FORCEnet and indicate how FORCEnet implementation measures relate to each of these concepts.**

**Recommendation for OPNAV, NETWARCOM, and MCCDC: Make clear that FORCEnet applies to the entire naval force and not just to its information infrastructure component. In so doing, the organizations should specifically indicate that the concepts of employment and the architectures devel-**

**oped must apply to the operation of the whole force and not just to its information infrastructure component.**

The committee finds the definition of FORCEnet quoted above to be adequate,[3] but notes that this definition implies three components:

- The doctrine, tactics, techniques, and procedures for conducting network-centric operations and the warriors trained in these concepts;
- The materiel developed and acquired in accordance with an architectural framework that enables these operations; and
- An information infrastructure that integrates the warriors and the materiel in the conduct of these operations.

Some equate FORCEnet with only the last of these three components. However, the committee believes that all three must be pursued in parallel, and uses the term "FORCEnet Information Infrastructure" (FnII) to refer to the third component. It will be important for the Navy and Marine Corps to accept the quoted definition of FORCEnet as being inclusive of the entire naval force and not just its information infrastructure, and thus to pursue concepts for employment of capabilities and architectures for the entire force.

### Accept That FORCEnet Has No Fixed End State

**Recommendation for the CNO and the Commandant of the Marine Corps (CMC): Promote as a guiding principle that the realization of FORCEnet capabilities will require a process of continuous evolution involving the close coordination and coupling of the individual departmental functional processes—operational concept and requirements development, program formulation and resource allocation, and acquisition and engineering execution.**

There is no defined end state for FORCEnet,[4] just as there are no defined end states for the Navy and Marine Corps. The realization of network-centric capabilities will require the coevolution of materiel and concepts for employment of that materiel, just as the Navy coevolved materiel and concepts in its development of sea-based air power and the Marine Corps coevolved materiel and concepts in its development of amphibious warfare capabilities. The full realization of these developments required the period of time between the world wars, and the full realization of network-centric capabilities will likely also occupy a generation.

---

[3]See also Section 2.2.
[4]See also Sections 4.6 and 7.3.

The coordination of concept development (including experimentation), requirements generation, program formulation, and program execution during a protracted coevolution, across the naval forces, will be a major challenge because there is no official below the Secretary of the Navy (SECNAV) who is responsible for all of these activities.

### Establish Governance Mechanisms That Span the Office of the Chief of Naval Operations, the Acquisition Community, and the Fleet

**Recommendation for the SECNAV, in conjunction with the CNO and the Assistant Secretary of the Navy for Research, Development, and Acquisition (ASN(RDA)): Develop a means to integrate more closely the Navy's program-formulation and acquisition functions, to ensure that adjustments in program execution are consistent with program intent and best serve the overall need of providing forcewide FORCEnet capability.**

**Recommendation for the CNO, in conjunction with the ASN(RDA): Establish a set of FORCEnet goals to be realized by specified dates in order to drive the implementation process.**

The committee notes with approval the action by the ASN(RDA) to establish an executive committee for overseeing and synchronizing FORCEnet materiel acquisition, although more participation by the fleet would be desirable. However, this committee struggled with the challenge of finding mechanisms to propose for coordinating the responsibilities of OPNAV for program formulation and resourcing with the responsibilities of the acquisition community and for setting short-term and mid-term goals for the entire FORCEnet enterprise.[5] Statutory requirements that program executive officers (PEOs) report directly to their Service acquisition executives (the Service acquisition executive for the Navy is the ASN(RDA)) prevent interposing a coordinator between the PEOs and the ASN(RDA). After much discussion, the committee sees two options for a coordinating mechanism:

- Establish a FORCEnet board co-chaired by the Vice Chief of Naval Operations (VCNO) and the ASN(RDA), or
- Appoint a senior officer as the director of FORCEnet, reporting directly to the VCNO and the ASN(RDA).

Some members of the committee were skeptical that a FORCEnet board would have the continuity to accomplish much, and instead favored the appointment of a forceful three-star or even four-star flag officer as the director of

---

[5]See also Sections 5.7 and 7.3.3.2.

FORCEnet. Others noted that a director of FORCEnet, having no line authority over the PEOs and perhaps none over the relevant Deputy Chiefs of Naval Operations (DCNOs)—the DCNO for Warfare Requirements and Programs (N6/N7) and the DCNO for Resources, Requirements, and Assessments (N8)—would need to have recommendations turned into commands by his or her seniors, just as the chief of staff of a board would. A board would have a dedicated staff that would meet daily, just as the staff of a FORCEnet director would.

In considering these options, the committee noted that neither of them would appropriately link the requirements and acquisition processes to the needs of the operational community. Accordingly, the committee discussed the possibility that, in parallel with either of these options, the CNO might establish measurable FORCEnet capability goals with required dates, as well as an annual FORCEnet master plan, and charter the Commander, Fleet Forces Command (CFFC), to monitor the accomplishment of these goals from the perspective of the fleet. These goals and plans would be useful regardless of the establishment of any new requirements-acquisition oversight office.[6]

The committee did not attempt to make analogous recommendations concerning the Marine Corps, primarily because MCCDC currently oversees and integrates Marine Corps concepts, requirements, and experimentation.

### Devote More Resources to Developing Operational Constructs

**Recommendation for NETWARCOM, and the Second and Third Fleets especially: Devote significantly more resources to concept development. The criticality of concept development to the overall realization of FORCEnet capabilities certainly requires this increase. The committee recommends that the CFFC determine whether the increased resources would come by reassigning personnel already assigned to the organizations or by request to the CNO for additional personnel.**

Within the Marine Corps, MCCDC has long been responsible for both concepts and requirements. The Navy quite recently entrusted both of these responsibilities to the CFFC. Although the committee applauds combining these responsibilities in one command, it notes that the CFFC has delegated these responsibilities to diverse operational agents, with the NETWARCOM responsible mostly for the FnII, the Second Fleet for Sea Strike and Sea Basing, and the Third Fleet for Sea Shield.[7,8] The committee found too few resources dedicated to concept development and, in the case of the Second and Third Fleets, to the formulation of requirements. Further, the division of responsibilities generates a

---

[6]See also Sections 4.8.6 and 7.3.3.2.

[7]See also Sections 4.7 and 7.3.3.2.

[8]Sea Strike, Sea Basing, and Sea Shield are the three pillars of Sea Power 21, the overall vision for transforming the Navy.

requirement for close coordination between the operational agents—a requirement that is not always met, perhaps because of the shortage of resources.

Accordingly, the committee believes that the CFFC would benefit from a
greater assignment of resources to its operational agents for concept development
and to the Second and Third Fleets for requirements formulation. Pacific Fleet
resources could also be brought to bear on this need. In addition, the committee
believes that NETWARCOM's FORCEnet concept development and experimentation role is slowly being extended beyond the FnII and applauds further extension of NETWARCOM's role in this regard.

## Base Resource Allocation Decisions on Packages
## That Reflect Network-Centric Operational Concepts

**Recommendation for the N6/N7 and N8: Develop resource-allocation meth-
ods directed at realizing forcewide FORCEnet capabilities. Instead of basing
the methods on the current Naval Capability Packages, the Navy should
instead use "packages" that inherently reflect network-centric operational
concepts.**

The N6/N7 and N8 are responsible for formulating and resourcing programs.
Their current, bottom-up approach is structured in such a way that each Naval
Capability Pillar (Sea Strike, Sea Basing, Sea Shield, and FORCEnet) formulates
resource recommendations, and the resulting, narrowly defined FORCEnet—that
is, essentially the FnII plus some, but not all, intelligence, surveillance, and
reconnaissance—competes for resources with the platforms and weapons that it
is supposed to empower.[9]

Accordingly, the committee expresses its concern that the current Navy resource-allocation process is not constructed on packages—such as FORCEnet
Engagement Packs—that reflect network-centric operational concepts.[10] In addition, while the Navy has sought improved modeling and simulation tools commensurate with the needs of network-centric operations, these efforts to date have
been less successful than is desirable. In particular, they have not fully included
the "fog of war," and their setup procedures are so lengthy that few full campaign
simulations can be conducted during each program assessment cycle.

## Strengthen Architectural Development
## and Systems Engineering Capabilities

**Recommendation for the CNO and the ASN(RDA): Designate the Naval Sea
Systems Command (NAVSEA), drawing on its open architecture experi-**

---

[9]See also Sections 4.5, 5.4, and 7.3.3.2.
[10]See also Sections 4.9.2 and 7.3.2.2.

ence, as having a major role in developing the FORCEnet architecture, particularly as pertains to its representation of invariant boundaries and the ability to allocate functionality.

**Recommendation for the ASN(RDA) with the support of the systems commands and the relevant PEOs (primarily the PEO for Command, Control, Communications, Computers, and Intelligence (C4I) and Space; and the PEO for Integrated Warfare Systems): Develop the capability necessary to effect FORCEnet systems engineering. Very high standards, commensurate with the challenge, should be set.**

Materiel must be specified, developed, and acquired in accordance with an overarching FORCEnet architecture.[11] The committee does not find that the draft *FORCEnet Architecture and Standards* developed by the Space and Naval Warfare Systems Command (SPAWAR)[12] provides optimal guidance for architecture development: Volume I primarily surveys potential FORCEnet components without venturing into functional allocation, and Volume II primarily directs interoperability standards for the FnII. The committee prefers the approach of the Open Architecture for Combat Systems developed by a NAVSEA initiative.[13] That approach reflects the ideas of invariant boundaries and functional partitioning that are required to engineer complex systems.

Evolving the complex system that is the transformed naval forces in accordance with the FORCEnet construct and architecture will require highly skilled people drawn from DON and industry. Not only must these personnel ensure that the evolving parts work together effectively, but they must also eliminate the possibility of catastrophic failure modes affecting the entire force. The integration and testing of new capabilities will require access to a facility analogous to the Navy's current Distributed Engineering Plant for combat systems. The committee does not believe that the plans for the Joint Distributed Engineering Plant have crystallized sufficiently for FORCEnet to rely on them. Instead, by extending its Distributed Engineering Plant to meet FORCEnet needs, the Navy could help the realization of the Joint Distributed Engineering Plant.[14]

---

[11]See also Sections 5.3.4 and 7.4.1.

[12]Space and Naval Warfare Systems Command. 2004. *FORCEnet Architecture and Standards, Volume I. Operational and Systems View*, Version 1.4, San Diego, California, April 30; Space and Naval Warfare Systems Command. 2004. *FORCEnet Architecture and Standards, Volume II, Technical View*, Version 1.4, San Diego, California, April 30.

[13]Naval Surface Warfare Center. 2003. *Open Architecture Functional Architecture Definition Document*, Version 2.0, November.

[14]See also Recommendations 21 and 22 in Chapter 7 and Sections 5.6 and 7.4.2.

## Strengthen the Naval Coupling to the Combatant Commanders

**Recommendation for the fleet commands and Marine Expeditionary Forces (MEFs): Build on current interactions with regional combatant commands in order to grow the relationship between naval and joint concept development and experimentation. This means ensuring both that naval concepts are properly embodied in joint concepts and that they reflect the needs of the joint concepts.**

**Recommendation for the N6/N7 and the Deputy Commandant of the Marine Corps for Plans, Policies, and Operations: Work to articulate clearly how FORCEnet capabilities pertain to joint operations and satisfy the needs of combatant commanders.**

The functional allocation that is part of FORCEnet systems engineering must eventually extend across all Services.[15] Combatant commanders must be able to compose their capabilities from resources supplied by all Services. The Joint Defense Capabilities Program envisions a process through which combatant commanders' expressions of needs drive the acquisition process. The Naval Services must provide capabilities that fulfill combatant commanders' needs and make sure that the relevance of these capabilities is understood. The committee expresses its concern that the Fleet commands and Marine Expeditionary Forces do not appear to be effectively feeding the needs of combatant commanders into the CFFC and MCCDC requirements processes. Further, the committee believes that OPNAV may need to more actively articulate naval programs' relevance to joint capabilities.

## Exploit Global Information Grid (GIG) Capabilities While Preparing to Fill GIG Gaps and Determining the Limits of Network-Centricity

**Recommendation for the N6/N7 and the Marine Corps Director for C4I: Adopt a prudent course with respect to joint GIG programs, endorsing the further development of these programs but also requiring a clear and continuing assessment of their technical and programmatic progress. In this context, the N6/N7 and the Director, C4I, should clearly understand the limits of applicability of network-centric capabilities, especially at the tactical level.**

Although FORCEnet is not a program, FORCEnet-related programs will have joint impact and therefore will entail joint certification. The joint network-centric information infrastructure is being planned by the Assistant Secretary of

---

[15]See also Recommendation 27 in Chapter 7 and Sections 3.3 and 7.5.2.

Defense for Networks and Information Integration (ASD(NII)) as the GIG. The components of the GIG are being developed by combat support agencies and by the Services. The FnII may be considered the maritime portion or extension of the GIG.[16]

By coordinating the development of the GIG, the ASD(NII) is enabling interoperability and focusing Service and agency efforts. However, the committee has some concerns about the GIG.

GIG backers have promised that communications bandwidth will no longer be a constraint on system design. Relying on this promise, the Defense Information Systems Agency has embraced a services-oriented enterprise information architecture that has numerous advantages, but that multiplies communications capacity and connectivity requirements and depends on continuous high-capacity, low-latency connectivity.

However, the Transformational Satellite Program that promises high capacity keeps being delayed. Even when it is completed, and even if the Navy develops and deploys suitable shipboard terminals, the Navy's communications capacity will not be infinite, and naval ships will still be subject to satellite communications interruptions caused principally by antenna blockage. Also, the GIG programs ignore the challenging problem of communicating with submarines at speed and depth in order to make them part of the networked force.

The challenge for DON will be to be prepared to exploit GIG capabilities as they come online, while pursuing science and technology to meet naval-unique challenges such as the antenna and submarine problems just cited as well as to address the information-management problems specific to naval operations. While the highest priority should be given to ensuring robust connectivity across naval units and to resolving naval information-management challenges, DON will also have to contribute to meeting challenges common to all network-centric operations, such as ensuring the security and reliability of mobile network infrastructure.

Despite the improved information infrastructure promised by the GIG, network performance or reliability may be insufficient for some functions. Mission-based, red-teamed analyses are needed to determine, for any given level of network capability, what functions are best performed locally rather than in a distributed fashion. The provision of alternate communication paths and facilities and opportunities for training in and rehearsal of operations suffering from degraded network capabilities must not be neglected.

---

[16]See also Recommendations 30 and 31 in Chapter 7 and Sections 3.6, 3.7, 5.4, 6.5.2, 7.5.1, and 7.5.2.

# 1

# Transforming the Navy and Marine Corps into a Network-Centric Force

Long before naval leaders began articulating network-centric warfare,[1] the U.S. Navy integrated weapons and sensors at diverse locations to perform its missions. In the earliest days of naval combat, flag signals were used to place ships into formations that permitted the concentration of their firepower. In the mid-20th century, antisubmarine warfare (ASW) operations depended on long-range but limited-accuracy sensors cueing an air platform to a point where it could deploy shorter-range but more-accurate sensors that could yield a targeting solution.

The timescales of these ASW operations permitted the use of voice and teletype person-to-person communications. The more time-compressed challenge of coordinated air defense against kamikaze aircraft motivated the development of the Naval Tactical Data System (NTDS), which used first-generation computers to exchange radar pictures from multiple ships to create a common picture for the air defense controller. The accelerating pace of computational capability has led to the vision of network-centric operations, which have been defined as

> . . . military operations that exploit state-of-the-art information and networking technology to integrate widely dispersed human decision makers, situational and targeting sensors, and forces and weapons into a highly adaptive, comprehensive system to achieve unprecedented mission effectiveness.[2]

---

[1]For example, see VADM Arthur K. Cebrowski, USN; and John J. Garstka, 1998, "Network Centric Warfare: Its Origin and Future," *U.S. Naval Institute Proceedings*, January, pp. 28-35.

[2]Naval Studies Board, National Research Council. 2000. *Network-Centric Naval Forces: A Transition Strategy for Enhancing Operational Capabilities*, National Academy Press, Washington, D.C., p. 1.

However, the full realization of network-centric operations presents technical, operational, and management challenges for which little historical guidance is available. The linkage of today's systems into network-centric forces will be an exceptionally large, complex undertaking; its technical aspects might be termed "complex system" engineering. The transformation it will bring to operations is so profound that the impacts cannot yet be fathomed.

The implementation of network-centric operations is unfolding in an uncertain environment. Navy and Marine Corps missions are in flux as the entire Department of Defense (DOD) undergoes a significant transformation. Perhaps more important, the pace of technological change has now increased so much that technology changes far faster than new naval systems can be designed and brought to the field. In the old days, the Navy could repeatedly field state-of-the-art devices and systems; today's systems are often obsolete before being fielded.

The operational challenge is to devise concepts of operation that exploit these technical capabilities as the United States moves from stovepiped, industrial age naval forces to a geographically dispersed, information age force that exploits all available information and seamlessly engages adversaries at the time and place of its choosing. The management challenge is to create mechanisms to coordinate the responses to the technical and operational challenges. Both the operational and the management challenge must be tackled before the Navy and Marine Corps can achieve the full promise of network-centric operations.

## 1.1 THE PROMISE OF NETWORK-CENTRIC OPERATIONS (A SCENARIO)

FORCEnet has broad applicability and promises to enable a wide variety of missions to be carried out with greater speed and effectiveness. This study committee—the Committee on the FORCEnet Implementation Strategy—decided that a vision of how FORCEnet could play out in the future in a specific, complex, joint scenario might illustrate the range of capabilities and the effectiveness that FORCEnet would enable: shared awareness, collaboration, responsive tasking, automated analysis and data synthesis, information composability, tactical decision support, collaboration and tasking of joint assets, force self-synchronization, rapid force composability, automatic incorporation of new sensors to form a new common picture, real-time composability of allied force response, and overall speed and decisiveness of command. The following is a scenario that mentions fictitious names and is set in the future; the shaded blocks highlight the capability illustrated.

It has now been 2 years since the political coup and consequent civil unrest in Camolia resulted in a small, U.S.-led peacekeeping mission being sent to the country. While the threat of civil war has been kept in check, in the past few months there has been a growing number of border strikes into refugee camps in neighboring Angeria, leading to concerns that Angeria would respond by striking Camolia and seizing its oil-producing regions. It is in this context that our story begins . . . .

*June 6, 2014:* Onboard the USS *Newport Beach*, a Joint Task Force command ship.

*0800:* MajGen John Gamble, USMC, commander of the 5,000-member peacekeeping force in Camolia, enters his ready room promptly at 0800 to begin the day's briefings. Little does he know what the next hour will bring.

Shared awareness: land tactical input to operational commander

*0802:* LCDR Smith, USN, commander of a small SEAL (Sea, Air, Land) unit outside the Camolian town of Rio del Agua, notes an unusually large truck convoy coming from Angeria. He launches a Dragon Eye unmanned air vehicle (UAV) to monitor the trucks. Reconnaissance data from the Dragon Eye are automatically transmitted directly to Commander Smith's personal digital assistant (PDA) and over the theater communications network (linked via a high-altitude UAV to the Transformational Satellite (TSAT) network) into the Joint Integrated Warfare Picture (JIWP). Commander Smith annotates the data with his concerns for General Gamble's staff onboard the *Newport Beach*.

Responsive tasking: operational to strategic

*0804:* General Gamble's staff patches in current DOD space-based radar (SBR) ground moving target indication (GMTI) and recent overhead imagery data to help determine the intent behind the trucks' movement. The track processing service (one of many such information-processing and advisory

capabilities embedded in the Global Information
Grid (GIG) infrastructure) quickly retraces the
GMTI track associated with the convoy to determine
the location of origin and correlates it with spot
imagery to confirm that the trucks originated from a
small Angerian military post and are likely troop
transports. General Gamble declares an end to the
morning briefings and orders an immediate determination of the trucks' likely destinations and the
potential force options for defeating Angerian occu-
pation of those locations.

Automatic
background
analysis and data
synthesis

*0806:* The nearly instantaneous response to General
Gamble's request is enabled by networks and enterprise services in the GIG infrastructure. The JIWP is
updated with detailed maps of the areas involved,
showing road networks, buildings, and critical
facilities, derived from the National Geospatial-
intelligence Agency (NGA) Digital Point Position
Database. General Gamble's intelligence officer
pulls up the NGA's trafficability and movement
analysis service to highlight likely destinations for
the invading force, with times of arrival. Simultaneously, the tactical alert service accesses the database on Angerian force capability and readiness
posture to display all potential ordnance sources—
aircraft, cruise missiles, surface attack missiles, and
ground forces—along with approximate time to
intercept and probable weapons effectiveness.

Information
composability

Decision-support
applications

*0807:* General Gamble directs LtCol Tucker, USMC,
his land force commander, to prepare a company of
Marines at the Dosama Airport for rapid deployment. He also directs that two Joint Strike Fighters
(JSFs) be readied for close air support. The resource
management service tasks two JSFs from the aircraft
carrier USS *Ronald Reagan*; assigns a weapons
loadout of the new 100 lb, guided Joint Direct
Attack Munitions (JDAMs); and downloads relevant
meteorological, terrain, and feature data to the
aircraft mission computers. As these actions are

Navy and land
force joint
operation

being carried out, the tactical alert service posts an advisory that Angerian submarine movement has been detected and that it may pose a threat to coalition force surface units. General Gamble asks where Angeria's submarines are.

**Responsive tasking of theater assets**

*0808:* The command center has received an alert that a surfaced submarine, being tracked by SBR since it left port, has apparently submerged. The theater's high-altitude, loitering UAV is retasked to confirm that the submarine has submerged and to determine whether Angeria's remaining three diesel submarines remain in port at Awafa. The resulting imagery displayed on the JIWP indicates that only one submarine is missing from the pier and not on the surface. Seeing this development, Commander Jones, onboard the littoral combat ship USS *Sea Sprite*, informs General Gamble that he is launching tactical UAVs to find and track the submerged submarine. Onboard the *Sea Sprite*, crews begin refitting two UAVs for submarine hunting; the UAVs will be refitted and airborne in 10 minutes. A search plan based on the submarine's last known location is developed by the ASW planning service and downloaded to the *Sea Sprite*'s command center and the UAV mission computers prior to launch.

**Self-synchronization**

*0811:* The SEAL unit's Dragon Eye imagery (relayed through the JIWP) shows that the trucks have forcibly entered an oil-processing facility outside Rio del Agua, less than a mile from the SEAL unit. Angerian troops with heavy weapons (including Swedish-built, tube-launched antiaircraft missiles) are seen attacking security forces at the facility. The image analysis service (supporting the JIWP) provides an estimate of troop strength, troop distribution, and weapons.

*0812:* General Gamble directs that a folder for "coalition eyes" containing convoy movement history and declassified Dragon Eye imagery of the attack be prepared in order to obtain timely approvals, as necessary, for further action.

*0814:* After assessing the options and ensuring that updates on the situation have gone to all friendly force locations in theater, General Gamble directs the Marine ready forces at Dosama to intercept the ground forces outside Rio del Agua—a distance of approximately 100 miles, with an estimated time of arrival (ETA) of 30 minutes in Ospreys. The Marine company uses its JIWP-Lite for en route planning and rehearsal aboard the Ospreys to prepare its plan of attack. The SEAL detachment is directed to make the best speed possible to an overlook position outside the oil facility and to provide enemy force positions and covering fire for the arrival of the Marines. The JSFs are launched and tasked to provide joint close-air support for the operation.

Rapid force composability

*0815:* Upon receiving its orders, a team of Marines and SEALs queries the NGA database and collaborates on the best vantage point from which to conduct its mission. Almost immediately, the NGA database returns a selection of three protected, elevated sites and quickest routes. Commander Smith elects to split his team into two groups and selects their routes. The JIWP-user (Marines and SEALs) PDAs and the involved aircraft systems are automatically updated with projected routes of all blue forces. The teams have an ETA of 10 minutes.

Ability to pull information globally from DOD sources

*0817:* With JSFs en route, the JIWP automatically updates its tactical picture with data from the superior JSF radar. Real-time Dragon Eye imagery analysis indicates that the Angerian forces have noticed the Dragon Eye and have launched a guided missile to intercept it . . . the Dragon Eye is lost.

Common picture, incorporating best current sensor inputs

*0818:* The *Sea Sprite*'s UAVs are launched and rapidly begin their assigned search pattern, laying a field of sonobuoys. Data from the sonobuoy field are

Seamless data with sensor tracks from theater and national systems incorporated into the JIWP to be fused with the ship's sonar data to provide an integrated subsurface picture.

Tactical decision support

*0820:* The JSFs are now on station over the oil facility. Having accessed the tactical data for the Angerians' weapons, the pilots have picked the minimum altitude that will be out of range of enemy weapons. Through the JIWP, the pilots track movements of Angerian vehicles around the facility.

*0825:* The team of SEALs reaches its position and sets up a ground imaging system to monitor the facility. With the SEAL team's updates, the JIWP now continuously tracks the motion of nearly every visible Angerian troop and relays it to the Marines. The SEAL team also readies its own guided missiles for use in support of the Marines' arrival.

Automatic incorporation of relevant new sensor data in the common picture

*0834:* The sonobuoy field has detected a likely submarine signature, target motion analysis is begun, and the UAVs are directed to begin a paired magnetic anomaly detection pattern to localize the target. If its location is confirmed, the submarine is nearing the British frigate HMS *Trincomalee*. The UAV data and target analysis are continuously accessible through the JIWP to the *Trincomalee* command center.

Off-board organic response enabled by real-time composability

*0838:* The submarine location is confirmed, and at nearly the same instant sonar indicates that a quiet torpedo has been launched toward the *Trincomalee*. Submarine and torpedo track information is supplied by the JIWP to tactical displays on the *Trincomalee* and *Sea Sprite*. The tactical alert service automatically provides courses of action consistent with rules of engagement, along with time constraints and probabilities of success. The tactical officers collaborate and rapidly determine that the *Trincomalee* will employ a torpedo-hunting missile in its own defense, and simultaneously, the *Sea Sprite* will deter, but not destroy, the submarine.

*0839:* The torpedo is successfully destroyed by the *Trincomalee*. A second torpedo-hunting missile is launched from the UAV and detonated within 30 meters of the submarine. The intent is to shake the submarine badly but not sink it; administration policy is to preserve the region's balance of power after a regime change in Angeria.

*0841:* The JSFs are each directed to launch two JDAMs at the Angerian trucks. Timing is coordinated through the JIWP so that the JDAM detonations will coincide with the imminent SEAL team attacks.

Collaboration using shared awareness

*0842:* The SEAL team, having received precise updates through the JIWP on the Marines' ETA, begins its missile assault on the enemy forces. JDAMs land at the same time. The Angerians return uncoordinated fire and are distracted from the Marines' arrival.

*0844:* The five Marine Corps Opsreys land on the opposite side of the facility from the SEAL team and engage the enemy ground forces. The Angerians realize that they are surrounded and outgunned. They surrender.

*0900:* General Gamble takes a deep breath and comments to his officers about how fast everything moved and how well it was coordinated. He thanks all those involved.

## 1.2 KEY CHARACTERISTICS NEEDED TO ACHIEVE THE PROMISE OF NETWORK-CENTRIC OPERATIONS

As illustrated in the preceding scenario, a fully networked force could potentially increase naval combat capabilities enormously. Network-centric operations will increase blue force tracking ability and decrease uncertainties and confusion, often termed the "fog of war," in turn increasing flexibility and options, accelerating decision making, and decreasing vulnerabilities. However, the force's information infrastructure will need a set of essential characteristics to make this possible:

- Robust availability,
- Assurance and trustworthiness,
- Coherence—avoidance of "Tower of Babel" problems,
- "Plug-and-play" composability of networked forces, and
- Adequate capacity and timeliness.

Each characteristic will be very challenging and indeed some are beyond the current state of the art. The following subsections briefly discuss these issues.

### 1.2.1 Robust Availability

Most fundamentally, it is imperative to recognize that a network-centric force is a network-dependent force. This reality imposes three requirements on the network, but also on the operations that employ it:

1. The network must be extremely robust, with sufficient redundancy to adapt to losses of component portions. In the past, large-scale distributed systems, such as networks and electrical systems, have often proved to be surprisingly fragile. Specific processes must be put in place for adapting to each potential loss.

2. Since the loss of a portion of the network is likely to reduce the network's capacity and capability, operations must quickly adjust to reductions in communications, and training must include operating at each level of reduced capability.

3. Since there is a significant likelihood that some platforms (e.g., ships or aircraft) will lose connectivity totally, they must possess sufficient local capability to allow effective operation in such circumstances.

### 1.2.2 Assurance and Trustworthiness

Military forces must be able to rely on the network and its information without constant concern that the information that they are using has been "poisoned" by deliberate acts of an adversary, or that an adversary is "inside" the network, or that the entire network might suddenly collapse under enemy attack.

Traditional military communications have emphasized security levels and cryptography to protect information. These forms of protection are still essential, but as the network grows ever larger, it is almost inevitable that some portion of it (e.g., sensor nodes or overrun ground vehicles) will at some point be controlled by an adversary. An adversary that can observe the common operational picture (COP) will have an advantage, and one that can "poison" data used by U.S. forces may cause long-term damage that is hard to find and undo.

Even worse, when the entire force is networked, adversaries could cause devastating, widespread effects within a single operation. One worm could take down communications in an entire theater, or even disable the communications of the worldwide assembly of U.S. forces, and it might take days to fully recover. This is a new level of threat for U.S. forces. It is deeply sobering but very true that the ongoing transformation to deeply network-centric operation opens the door to far more serious vulnerabilities for naval forces than they have faced historically.

### 1.2.3 Coherence—Avoidance of "Tower of Babel" Problems

One clear benefit of network-centric operation is a common operational picture through which operators can see at a glance their own current locations, the positions of nearby friendly forces, and current estimates of enemy locations. Unfortunately, experience has shown that such "common" pictures are anything but common. The committee heard first-person anecdotes of compelling graphic displays that were completely wrong, with many of their icons showing incorrect or outdated position information, thus making the entire picture worse than useless. Another anecdote told of 21 unrelated "common" operational pictures.

In addition to having an accurate common operational picture, it is equally desirable to share sets of radar tracks (e.g., for aircraft over a theater) to which a number of different sensors each contributes. However, experiments at the exercises of the All Services Combat Identification Evaluation Team have shown for many years that the DOD is nowhere near being able to properly correlate tracks contributed from different sensor systems. Instead, a single physical object may be represented as many different tracks at different locations because of the inherent inaccuracies of the individual sensors.

Three important challenges underlie the ability to avoid this kind of confusion: the creation of interoperable data definitions, the development of open systems for information dissemination, and the formulation of information services that enable mathematically consistent processing of data and information. Each is a hard problem, combining both technical and cross-organizational programmatic difficulties.

### 1.2.4 "Plug-and-Play" Composability of Networked Forces

Today's system engineering builds a reliable system by bounding the problem and decomposing the larger system into a set of smaller subsystems with their own derived requirements. These subsystems can then be assembled into the overall system with good assurance that the result will meet its goals for reliability, timeliness, accuracy, and so on.

This successful technique cannot readily be applied to systems that are assembled "on the fly," however. It is one thing to engineer a Cooperative Engagement Capability system, which is a complex and highly successful distributed system. It is quite another thing to plug together a previously unrelated set of sensor systems and weapons in the field and expect such an arrangement to work. Thus, tension currently exists between the desires of commanders for the "plug-and-play" interoperability of forces and their information systems, and the ability of system designers to produce systems that will work reliably when lashed together.

### 1.2.5 Adequate Capacity and Timeliness

Finally, at the most basic levels, the network infrastructure must provide adequate bandwidth, and it must deliver information in a sufficiently timely manner that it is still useful when it arrives. This will likely prove challenging for the Navy and Marine Corps even when new satellite systems and peer-to-peer radio networks are in place, since connectivity to mobile platforms is by its nature slower and less reliable than that to fixed sites. The Marines have a harder problem in this respect than the Navy has, since closing a radio link to a small terrestrial vehicle or to a dismounted Marine is very challenging.

As a concrete example of where the Navy stands today, the great majority of ships have only 32 kilobits per second (kb/s) of bandwidth for network connectivity. Thus a fighting ship often receives far less bandwidth than a home personal computer does. Furthermore, this connectivity may only be available 70 percent of the time. These capabilities are hardly compatible with a "fully networked fighting force."

Timeliness is also a key issue. In recent years, tactical communications have been roughly divided into messages with time requirements measured in hours or minutes (e.g., the formulation of a ground attack plan and issuance of the execution order), in seconds (battle management), and in hundreds of milliseconds (fire control). General-purpose networks, based on Internet technology, already convey the first type of messages, and they could convey the second with a modest system engineering effort. However, it may still be too early to tackle the hard, real-time, weapons-control control loop in a general-purpose network.

### 1.3 "ENGINEERING THE VISION"

Although it may not be immediately apparent, the information infrastructure needed to support the network-centric operational vision is in fundamental ways unique to the military. Analogies with the Internet may be illuminating, but the infrastructure needed by the military cannot be achieved by simply purchasing and plugging together commercial systems. Its development will be a very large scale, long-term, and highly technical undertaking marked by the following characteristics:

• *Unprecedented scope.* The military requires a worldwide, always-available system that provides high-quality, protected connectivity anywhere in the world on little or no notice. This connectivity must be provided everywhere, from the depths of the ocean to the centers of foreign cities. On the face of it, this is as large an undertaking as any tackled by companies such as Verizon or AT&T, which each devotes hundreds of thousands of employees and tens of billions of dollars per year to maintain and operate its networks.

• *Unprecedented need for robustness.* The military's information infrastructure must also be designed to withstand various levels of attack—not just the annoyances created, for example, by teenage hackers, but the heavy attacks launched by determined nations with significant budgets and top-notch technical expertise. Potential attacks range from the old-fashioned jamming of satellite links, to the use of electromagnetic pulses to disable commercial computers, to the deliberate "poisoning" of significant information within U.S. military databases, to the launching of network viruses and worms. No commercial systems are designed to withstand this range of threats.

• *Significant difficulties in execution.* Finally, this system must be procured and constructed within DOD's legal and organizational frameworks. Such a large system necessarily cuts across tens or hundreds of procurement programs, bringing a high likelihood of uncoordinated and incompatible development. Systems engineering of a very high order will be required, but at present the Department of the Navy (DON), and indeed the DOD as a whole, possesses no great depth of engineering talent. While there are excellent software and systems engineers in the enterprise, there are too few of them.

Given these observations, one might ask whether DON should even try to proceed with a transformation to network-centric operations. In the committee's view, the answer is a resounding and unanimous "Yes." In fact, this transformation is already well underway and is very highly desirable. The question, then, is not *whether* to proceed in the face of such significant challenges, but rather *how*.

As discussed in this report, the committee believes that the best strategy going forward is to tackle the problem little by little, with an emphasis on near-term warfighting capability. It would be fruitless to try to draw up a detailed plan when tackling such a large problem. Instead, the Navy and the Marine Corps should perform a rapid, focused, spiral evolution of technology and operational concepts, working out the easiest problems first and deferring the hardest ones. The one exception is information assurance, which is very difficult but so critically important that it must be addressed immediately and continuously. A few early successes will help maintain the proper momentum and fuel the process by which operations and technology can coevolve.

In short, the Navy and the Marine Corps would be well advised to treat this problem of "engineering the vision" as one of the largest undertakings in their history and to deal with it appropriately. It would be a serious mistake to underestimate the scope of the effort that will be required.

## 1.4  WHERE ARE WE TODAY?

The Navy and Marine Corps are already partway down the path toward network-centric operations, as indeed are the joint forces as a whole. In fact, in some areas the Navy has been performing network-centric operations for decades. One striking example is ASW, by which a set of platform sensors with very limited range can, when efficiently coordinated, find difficult targets in large areas. But in a broader context, many of today's missions now exploit network connectivity. It goes without saying that these operations almost always involve deeply joint efforts.

This section briefly discusses how network-centric the recent Navy and Marine combat activities have been and which programs, already underway, are starting to build out the first major network-centric capabilities for use in future naval operations.

### 1.4.1  Recent Navy Operations

Perhaps the most striking aspect of recent Navy operations has been the dramatically shortened Air Tasking Order (ATO) cycle and the changing relationship between targeting and the ATO. The ATO cycle has decreased from 72 to 24 hours, and during the conflict in Afghanistan, 80 percent of the targets destroyed were passed to pilots after they had left the carrier deck. A key element in this success has been digital links between forward air controllers and aircraft.

Logistics has also greatly improved. With maritime prepositioning, transporting equipment by ship, and using C-17s, the Navy delivered four times the tonnage of goods and equipment for Operation Iraqi Freedom (OIF) that it delivered for the earlier Desert Storm, and in 4 months instead of 7. It is also apparent that the traditional two-carriers-at-sea rotation did not hold up, as seven carriers supported OIF.

Navy assets for OIF were bandwidth-limited, even with a remarkable surge in commercial satellite augmentation, for a variety of reasons. Some were purely technical—for example, small ships had no choice but International Marine/ Maritime Satellite (INMARSAT) connectivity, which resulted in very low bandwidth (32 kb/s maximum) with very poor availability (about 70 percent). Other reasons were more operational—for example, the ground forces were allocated a higher fraction of available military bandwidth for what was, after all, primarily a ground fight.

### 1.4.2  Recent Marine Corps Operations

During OIF, communications requirements of the Marines relied primarily on legacy systems that have been used for a number of years. Line-of-sight Single-Channel Ground-Air Radio System radios, squad handheld radios, and some high-frequency and ultrahigh-frequency satellite terminals were the principal items used. Communications down to the battalion level were fairly reliable, though imperfect below that level. As the attack progressed, radio communications between adjacent units enabled small-unit leaders to coordinate actions and speed up the advance. Blue force tracking, for most units, continued to be accomplished through unit boundaries.

As for battlefield visualization, some division and regimental units had a rudimentary tactical operational picture (TOP). The information displayed by the TOP was not viewed with confidence by the commanders or staffs. Information currency and service connectivity were the most frequent concerns. The equipment and the operators were not able to adjust to the rapid advance and tracking of so many units for such distances over such a large area. The division commander, MajGen James N. Mattis, USMC, related to the committee that the COP did not really contribute to the battle synchronization.[3] He indicated that a reliable, accurate COP would be helpful at major headquarters, but that current systems do not provide the connectivity or reliability required by small units constantly on the move.

### 1.4.3  Basic Infrastructure—A Common Information Technology Infrastructure Across the Force

In recent years, the Navy and Marine Corps have installed a solid, almost ubiquitous information technology (IT) infrastructure built from standardized commercial computers and networks. The Navy-Marine Corps Intranet (NMCI) provides standardized IT services in the United States, while the Navy IT program Information Technology for the 21st Century (IT-21) to improve shipboard communications and computing capability and the Marine Corps Enterprise Network (MCEN) provide similar services to deployed forces of the Navy and Marine Corps, respectively. These programs have provided an essential first step toward network-centric operation.

---

[3]MajGen James N. Mattis, USMC, Commanding General, 1st Marine Division, presentation to the committee on December 16, 2003, Camp Pendleton, California.

NMCI provides the required homogeneity to the transport layer of operations to support information sharing and reach-back to Navy and Marine Corps shore-based infrastructure, and IT-21 and MCEN provide the same to the Navy and Marine Corps fighting units. None of these initiatives directly addressed applications interoperability except for that involving basic office functionality; however, their existence is essential to achieving data and applications interoperability. The interoperability of legacy applications with enterprise-level security policy is the biggest problem that the NMCI has faced. The same issue will be a challenge for all network-centric systems going forward.

### 1.4.4 Basic Infrastructure—The Global Information Grid

Even more recently than the installation of the NMCI, an energetic effort has been launched to design and build the key technological capabilities required to link all tactical and strategic forces into a unified GIG. The Assistant Secretary of Defense for Networks and Information Integration (ASD(NII)) of the Office of the Secretary of Defense (OSD) has led this effort. In this committee's view, this OSD-led effort has been focused and extremely well conducted to date.

Basic connectivity will be implemented by a set of related programs. The GIG-Bandwidth Expansion (GIG-BE) program will bring high-speed fiber connectivity to bases worldwide. The Transformational Communications Architecture (TCA) will provide robust, high-capacity satellite connectivity to forces in the field. The networking aspects of the Joint Tactical Radio System (JTRS) will extend tactical connectivity with mobile, ad hoc networks. The High Assurance Internet Protocol Encryptor (HAIPE) program is introducing modern, high-speed cryptographic services. All of these programs share a common technical architecture based on next-generation Internet Protocol (IP) technology, IP version 6 (IPv6). These programs are exceptionally important for the Navy and Marine Corps, as they will provide the basic network connectivity for military forces.

Newer ASD(NII) programs aim to provide network services beyond bare connectivity. Among them is the Network-Centric Enterprise Services (NCES) program, which is planned to provide a standardized layer of network services that can be employed by all military-specific applications programs across the DOD. It is still too early to say if these programs will be as coherent and promising as the connectivity programs.

### 1.5 ADDRESSING THE CHALLENGES

As exemplified above, the Navy and Marine Corps are already on a path toward networked operations. However, as previously noted, the full-scale transformation to network-centric operations will be a large and difficult undertaking with many impediments. This report considers each major impediment and makes specific recommendations for addressing each of them. This section briefly outlines these challenges and indicates the chapter in which each is discussed at length.

In addressing these impediments, the report is quite broad and general in scope. It is necessary to make clear what is not within this scope, given the particular charge in the terms of reference (see Chapter 8). The report does not consider the specifics of individual missions, be they traditional combat missions, such as strike and antiair warfare, or the "less regular" missions, such as combating terrorism and conducting stability operations. The perspective of the study is that a FORCEnet implementation strategy will lead to a set of capabilities applicable across all missions.

The terms of reference do not raise issues of coalition operations (although they do include joint operations), nor do they single out specific functional areas (e.g., training, logistics). Hence, these topics are not explicitly treated in the report in any great detail. Lastly, the report does not consider the cost implications of realizing FORCEnet capabilities. All of these are clearly important factors that would have to be considered in more detailed FORCEnet planning.

### 1.5.1 Unprecedented Scope and the Need for Common Understanding (Chapter 2)

Today, many communities are working on various components of the technical infrastructure needed for network-centric operations, but without much direct communication and interaction. Achieving the full technical and operational vision of network-centric operations will require some form of common, high-level coordination to ensure success. Warfare systems are still circumscribed, and their connection to the large, networked infrastructure remains unclear. A common understanding of both individual and shared objectives is paramount to making progress toward the vision.

### 1.5.2 An Evolving Joint Community (Chapter 3)

There is no doubt that nearly all network-centric operations, together with the information infrastructure that supports them, will be fully joint. Thus, the naval aspects of these operations and of this infrastructure can be considered only within the larger, joint context. However, this context is in a remarkable state of flux, with no end in sight. Major changes in the operational spheres, such as the creation of the U.S. Joint Forces Command (JFCOM) and the U.S. Northern Command (NORTHCOM) and their subsequent assumption of major duties in joint evolution, have been balanced by a thoroughgoing renovation in the area of joint programmatics. Whatever one might think about "transformation" as a warfighting concept, it is most strikingly a reality when it comes to DOD organizational structures and processes.

Two aspects of joint development deserve special comment. First, naval aspects of the network-centric infrastructure will be strongly shaped by the GIG and by programs emerging from OSD. It is inconceivable that naval information systems will not form part of this rapidly evolving, overall joint information infrastructure, in order to share information freely with the other Services and with national agencies. Second, experimentation will be a critically important tool for the evolution of network-centric operations and their supporting information infrastructure, and naval network-centric experimentation will take place within the broader context of joint experimentation, which in turn is rapidly evolving.

### 1.5.3 Coevolution of FORCEnet Operational Concepts and Materiel (Chapter 4)

To make FORCEnet a reality, a comprehensive approach requires more than exploiting current and emerging technologies in order to build and operate a network for warfighting. Because the introduction of FORCEnet capability will produce a major transformation in the conduct of naval operations, discovering nonmateriel solutions for meeting capability needs is as important as finding materiel solutions. Achieving this transformational capability depends on establishing processes that create interactions between the fielding of new technology and the development of new operating concepts.

New operational concepts are developed or evolve as a matter of necessity either from the introduction of new, improved capability created by new technology or as a change in the operational environment occurs. FORCEnet will require an iterative process of discovery in order to foster the development of operational concepts to take advantage of new technologies, or to highlight shortfalls in needed capability to stimulate and inform further technology development. The required coevolution is more than the spiral development of materiel to achieve a fixed performance goal. What is needed is an organized, integrated, dual-spiral process by which advancing technologies inspire new concepts and advancing concepts drive new technology investments in a mutually reinforcing way.

In the network-centric vision, large numbers of different systems, interconnected by the network infrastructure, are operating in a unified manner as a system of systems. In order for this to happen, the individual systems cannot be acquired independently, but must be designed, developed, tested, and fielded in a coordinated manner across the enterprise. Today, systems are acquired independently of one another; program managers are responsible for meeting their requirements independently of the success or failure of other programs. The unprecedented scope of network-centric systems—including sensors, networks, command and control, platforms, and weapons—demands a new management approach to system acquisition that subordinates the individual programs to an overall capability.

The current acquisition process has primarily been developed for and applied to the procurement of hardware and services, most often with the desired product and outcome specified in detail. This arrangement has permitted the establishment of in-process metrics to measure progress in performance and to minimize risks. The acquisition of network capabilities, whether through the integration of disparate existing systems, new capabilities, or other combinations, will require vastly different expectations. It will differ considerably from the traditional sequence of research and development (R&D), engineering development, limited production, and finally, production. Speed to capability may be an uncomfortable concept initially, since its implementation imposes a degree of process concurrency and risk taking that is currently minimized.

### 1.5.4 Engineering the Complex System (Chapter 5)

The network-centric Navy and Marine Corps that FORCEnet strives to create has all the attributes of a complex system.[4] Such systems are not only large and complicated, but are characterized by complex interactions among heterogeneous building blocks that adapt or are replaced over time as a consequence of environmental changes, leading to emergent behavior of the overall system. Even the FORCEnet Information Infrastructure (FnII) qualifies as a complex system.

Complex systems cannot be engineered by the traditional reductionist approach of partitioning fixed requirements among subsystems, each of which has a fixed and known behavior. This is clear with regard to network-centric operations because there is no fixed requirement for the network-centric Naval Services and because the components will be evolving. Instead, highly experienced engineers of large systems will be required, together with new approaches and tools. It will be essential that there be system engineering of portions of the materiel parts of the complex system, in the context of well-designed boundaries, and the use of an extension of the distributed engineering plant for multiple purposes from concept formulation through risk assessment and capability verification.

### 1.5.5 Technological Shortfalls (Chapter 6)

Some technical capabilities that will be needed for achieving the long-range vision of network-centric operations simply are not available today and may well not be developed in the commercial markets because they are too closely related to military needs. A brief catalog of these technical shortfalls makes the scope of this issue clear.

---

[4]Journals such as *Complex Systems* (ISSN 0891-2513, www.complex-systems.com, accessed July 24, 2004) and *Advances in Complex Systems* (ISSN 0219-5259, www.worldscinet.com/acs/acs.shtml, accessed July 24, 2004) explore the nature of these systems.

• Basic connectivity for platforms afloat generally relies on one satellite link or a small set of links per platform. Access to these links may be easy to deny in the future, and there is no immediately available alternative. Peer-to-peer, ad hoc networks between ships and aircraft might help solve the problem, but such technologies are currently immature. This problem is even more pronounced for platforms such as submarines and for almost all Marines and special operations forces.

• Information assurance is of the highest importance, and yet the current state of the art is not adequate to guarantee the requisite levels of assurance. Since information assurance is critical to network-centric operations, this area will require significant and sustained effort over the coming years, both within the Navy and in concert with related activities elsewhere in the Services and in the U.S. government.

• Information management and dissemination are still poorly understood, particularly when forces are composed as situations evolve. Well-defined methods for composing large software systems "on the fly" are currently not well understood, nor is there any good way to predict the behavior of the systems thus assembled.

• Large-scale modeling and simulation will likely be essential for the proper understanding and analysis of tomorrow's networked forces, but current technologies will probably not scale adequately. With current simulators, an exploration of network behavior alone can take weeks of real time for a moderately sized mobile network.

• Automated situational awareness with information fusion and user-defined visualization will be essential to distill needed information for specific users from the large volume of data traversing the network.

### 1.5.6 Navy and Marine Corps Cultural Issues (Overarching)

Three cultural issues may delay the transition to fully network-centric operations. The first involves the fact that the transition will require significant investments in information infrastructure, investments that will inevitably compete with those in weapons and weapons-delivery platforms. In a warrior culture, weapons, platforms, and their users command more respect than do computers, radios, and their users. A senior officer once remarked that "the volume entitled *Famous Naval Communicators* is thin indeed."

The second issue arises from the maritime tradition that the captain of a ship is "Master under God." Until radios were invented, a naval expeditionary force sailed with orders but was free to interpret them in accordance with the tactical situation. Network technology will give tactical commanders the situational awareness to self-synchronize without the feared and hated "rudder orders from above." However, some may fear that their seniors will second-guess their decisions or waste time with demands for explanations.

The third issue arises because a commander trusts most those forces under his or her command. Being responsible for the mission outcome and troops' welfare, every commander will try to plan for all contingencies. In most instances, the better the commander can control the situation, the better his or her chances of success. Dependency on capabilities that are provided by others adds uncertainty, risk, and worry. Yet network-centricity implies reliance on others who may be far away and belong to different communities.

## 1.6 FINDINGS

This chapter presents the committee's broad findings and observations on the problems of transitioning to network-centric operations without specific recommendations. Subsequent chapters investigate these issues in detail and provide concrete recommendations on specific actions that naval leadership can take. The broad findings and observations are as follows:

• Building the technical infrastructure for network-centric operations is an exceptionally large, complex undertaking in a very uncertain environment. A systems engineering perspective must be adopted up-front, with consideration given to issues of interoperability, security, reliability, availability, and the impact of network-centric operations on and from legacy systems.

• Rapid spiral evolution (that is, aiming for speed to capability) is generally more effective than drawing up a grand plan, and efforts should be directed toward clearly visible, near-term gains that are useful across many scenarios.

• Systems analysis and systems engineering will be essential to avoid chaos, dead ends, and parts that do not mesh into a whole.

• Even with great care to keep this from happening, network-centric operations will introduce large, new vulnerabilities to the Navy and the Marine Corps. This area should be a key focus for systems analysis and systems engineering.

# 2

# Defining FORCEnet

## 2.1 THE ORIGIN AND DEVELOPMENT
## OF THE FORCENet CONCEPT

### 2.1.1 ... *From the Sea* to Sea Power 21

For the U.S. Navy, the key transformation of the post–Cold War epoch was introduced in the 1992 white paper entitled ... *From the Sea*. That Navy–Marine Corps white paper embodied "a fundamental shift away from open-ocean warfighting on the sea toward joint operations from the sea"[1] and implicitly recognized that sea control was a means to an end—namely, the projection of power ashore. Accordingly, the white paper outlined an expeditionary force that could be "swift to respond on short notice to crises," "structured to build power from the sea," "able to sustain support for long-term operations," and "unrestricted by the need for transit or overflight approval."[2]

... *From the Sea* was quickly followed, in 1994, by another Navy–Marine Corps white paper, *Forward ... From the Sea*, which expanded the original document to include peacetime operations and conventional deterrence,[3] and in 1996

---

[1]Department of the Navy. 1992. ... *From the Sea*, U.S. Government Printing Office, Washington, D.C., September, p. 2.

[2]Department of the Navy. 1992. ... *From the Sea*, U.S. Government Printing Office, Washington, D.C., September, pp. 2-3.

[3]Department of the Navy. 1994. *Forward ... From the Sea: Continuing the Preparation of the Naval Services for the 21st Century*, U.S. Government Printing Office, Washington, D.C., September 19, p. 2.

by the Marine Corps's *Operational Maneuver from the Sea*,[4] explaining how the Marines proposed to execute this expeditionary concept. These were succeeded by a 1997 vision statement from the Chief of Naval Operations (CNO) entitled "Anytime, Anywhere," which defined the Navy role as being able "to influence events ashore, directly and decisively, from the sea, anytime, anywhere."[5] The vision also defined a broad littoral that encompassed "most of the earth's land masses, more than 80 percent of its population, and most of its capitals and major cities" and foresaw a mounting "area denial" challenge.[6]

The expanded role of naval forces in projecting power ashore that was outlined by the naval Services between 1992 and 1997 confronted head-on the historic mismatch between the limited combat power that could be projected from seaborne forces and the far larger assets that could ultimately be mounted from shore. It appeared to fly in the face of Admiral Nelson's adage, "A ship's a fool that fights a fort."[7] The new role, therefore, revolved about the challenge of giving "a highly trained, well-equipped, but perhaps smaller military force such as ours an impact so disproportionate to its numbers as to make it decisive in peace and in war."[8] This challenge, in turn, posed questions as to how much power of what kinds could be projected from sea-based forces and, by extension, how the impact of that power might be focused and multiplied—exactly the kinds of questions that were to become key joint issues in the wake of the events of September 11, 2001 (9/11) and OIF.

Starting in late 1997, this problem was posed to a succession of CNO Strategic Studies Groups (SSGs)[9] as well as to a committee of the Naval Studies Board of the National Research Council.[10] Highlights of these groups' contributions include the following:

• The first of the SSGs to address the problem—SSG 17—examined ways to better project power from the sea both by designing forces that could operate

---

[4]Headquarters, U.S. Marine Corps. 1996. *Operational Maneuver from the Sea*, U.S. Government Printing Office, Washington, D.C., January 4.

[5]ADM Jay Johnson, USN. 1997. "Anytime, Anywhere," *U.S. Naval Institute Proceedings*, November, pp. 48-50.

[6]ADM Jay Johnson, USN. 1997. "Anytime, Anywhere," *U.S. Naval Institute Proceedings*, November, pp. 48-50.

[7]Admiral Horatio Nelson (1758-1805).

[8]ADM Jay Johnson, USN. 1997. "Anytime, Anywhere," *U.S. Naval Institute Proceedings*, November, p. 50.

[9]The Strategic Studies Group is a body of from 8 to 13 fellows specially selected by the Chief of Naval Operations, the Commandant of the Marine Corps, and the Commandant of the Coast Guard to study a specific topic, tasked by the CNO for a 1-year fellowship. They work at the Naval War College under the guidance of a retired four-star admiral and are assisted by associate fellows drawn from the classes of the War College and the Naval Postgraduate School and a staff of qualified analysts.

[10]Naval Studies Board, National Research Council. 2000. *Network-Centric Naval Forces: A Transition Strategy for Enhancing Operational Capabilities*, National Academy Press, Washington, D.C.

without the use of local ports and airfields and by improving command and battlespace knowledge so as to multiply the power of sea-based forces through precise nodal targeting (in which timed firepower is focused on targets selected to have the greatest impact).[11]

• SSG 18 expanded upon this effort and introduced what was termed Sea Strike, a concept for greatly increasing the volume and precision of sea-based firepower in the conduct of joint operations. To that end, this studies group drew heavily upon network-centric warfare concepts to increase and focus the flow of information to commanders and sea-based strike forces including Marines as part of a joint response.[12]

• SSG 19 carried these network-centric solutions another step, with proposals for fully networking the naval Services into what was termed FORCEnet, which was to be an integral part of a larger and also fully networked joint force. It also proposed extending this structure to include a responsive expeditionary sensor grid.[13]

• SSGs 20 and 21 then refined the concept and addressed how FORCEnet might be implemented, as well as how the Navy might select, educate, and train the 21st-century warriors who would operate the new networked Navy.[14]

• In parallel with these activities, the CNO asked the Naval Studies Board to specifically examine a transition strategy for enhancing the operational effectiveness of naval forces through the application of network-centric operations. One result of the study carried out in response to that request is the following definition of network-centric operations:

> [Network-centric operations are] military operations that exploit state-of-the-art information and networking technology to integrate widely dispersed human decision makers, situational and targeting sensors, and forces and weapons into a highly adaptive, comprehensive system to achieve unprecedented mission effectiveness.[15]

The efforts summarized above—beginning with the . . . *From the Sea* white papers and continuing on through the efforts of five different SSGs and a CNO-requested study by the Naval Studies Board—are significant both because they

---

[11]ADM James R. Hogg, USN (Ret.), Director, CNO Strategic Studies Group, personal communication, November 9, 2005.

[12]ADM James R. Hogg, USN (Ret.), Director, CNO Strategic Studies Group, personal communication, November 9, 2005.

[13]ADM James R. Hogg, USN (Ret.), Director, CNO Strategic Studies Group, personal communication, November 9, 2005.

[14]ADM James R. Hogg, USN (Ret.), Director, CNO Strategic Studies Group, personal communication, November 9, 2005.

[15]Naval Studies Board, National Research Council. 2000. *Network-Centric Naval Forces: A Transition Strategy for Enhancing Operational Capabilities,* National Academy Press, Washington, D.C., p. 1.

represent a revolutionary change from the naval Services' traditional approach to naval warfare and because they embody a coherent direction in Navy thinking maintained through the tenure of two different Chiefs of Naval Operations and three different Commandants of the Marine Corps (CMC). These efforts also provide clear antecedents for ideas contained in the Navy's current Sea Power 21 vision and in the Marine Corps Expeditionary Maneuver Warfare (EMW) vision as well as for potential naval roles in much of the currently emerging Joint Operating Concept. Most of all, however, they provide a clear context for FORCEnet and what it is to accomplish.

### 2.1.2 Sea Power 21: The Dimensions of the FORCEnet Challenge

The Sea Power 21 vision of naval forces in the 21st century applies the sustained, decade-long evolution of Navy and Marine Corps thinking described above to a post–9/11 security environment "fraught with profound dangers: nations poised for conflict in key regions, widely dispersed and well-funded terrorist and criminal organizations, and failed states."[16] This environment, the Sea Power 21 vision contends,

> will produce frequent crises, often with little warning of timing, size, location, or intensity. Associated threats will be varied and deadly, including weapons of mass destruction, conventional warfare, and widespread terrorism. Future enemies will attempt to deny us access to critical areas of the world, threaten vital friends and interests overseas, and even try to conduct further attacks against the American homeland. These threats will pose increasingly complex challenges to national security and future warfighting.[17]

To provide the needed "Joint Capabilities"[18] to deal with this new, still-changing, and complex security environment, Admiral Vern Clark, Chief of Naval Operations, proposed a Navy vision that rests on three pillars: Sea Basing, Sea Strike, and Sea Shield. These Navy pillars are supplemented by the Marine Corps's EMW, with all of the operational constructs enabled by the network-centric-operations thinking and capabilities embodied in FORCEnet and, by extension, the still-evolving overall joint network. FORCEnet, Sea Basing, Sea Strike, and Sea Shield are, in turn, supported by three additional concepts: Sea Trial, Sea Warrior, and Sea Enterprise, which will provide the development, personnel, and acquisition underpinnings to carry out the Sea Power 21 initiative.

---

[16]ADM Vern Clark, USN. 2002. "Sea Power 21 Series, Part I: Projecting Decisive Joint Capabilities," *U.S. Naval Institute Proceedings*, October, p. 1.

[17]ADM Vern Clark, USN. 2002. "Sea Power 21 Series, Part I: Projecting Decisive Joint Capabilities," *U.S. Naval Institute Proceedings*, October, p. 3.

[18]ADM Vern Clark, USN. 2002. "Sea Power 21 Series, Part I: Projecting Decisive Joint Capabilities," *U.S. Naval Institute Proceedings*, October, p. 1.

## 2.1.2.1 The Three Pillars: Sea Basing, Sea Strike, and Sea Shield

*Sea Basing.* Sea Basing will provide the increased seaborne ability to protect, project, and support forces that was intimated in the . . . *From the Sea* white papers. The sea base is not a place or a unit but an assembly of capabilities that expands and contracts to match the requirements of the joint forces commander. Naval forces operating from dispersed locations using networked command-and-control structures will interface with naval shore facilities and set up strategic pipelines to support joint forces. As conditions change, the sea base provides the joint commander the ability to reconstitute forces at sea and redeploy them to exploit opportunities.

*Sea Strike.* Sea Strike will enable the Joint Force Commander to project decisive offensive power from the sea base. The projection of offensive power will be through the delivery of joint fires with increased range, lethality, accuracy, and timeliness from aircraft, ships, submarines, unmanned vehicles, and ground forces. Improved strike operations will be enabled by FORCEnet through the fusing of information from naval, joint, national, and multinational sensors and other sources. FORCEnet will create information networks with new levels of connectivity and integration, which will provide common and constant data throughout the force so as to permit offensive operations at the time and location of the Joint Force Commander's choosing.

*Sea Shield.* Sea Shield will produce an integrated, layered global defensive posture for joint forces operating in the littorals and at sea. Upon arrival in a region, naval forces will dominate the region's air, surface, subsurface, and cyberspace environments. Naval forces will provide this sustainable protective posture through a networked, distributed force that includes the deployment of air and missile defensive capabilities as well as surface, subsurface, land, and mine countermeasure assets.

## 2.1.2.2 Marine Corps Overarching Concepts of Operations

*Expeditionary Maneuver Warfare.* Paralleling the Navy's Sea Power 21 pillars is the Marine Corps concept of EMW. This concept describes how the Marine Corps will conduct operations within the complex, post–9/11 environment—a shift from reliance on the quantitative characteristics of warfare (mass and volume) to a realization of the importance of qualitative factors (speed, stealth, precision, and sustainability). Operating from the sea, Marines will maneuver to achieve decisive effects, concentrating forces at critical points to achieve surprise, shock, and momentum.

### 2.1.2.3 The Enabler—FORCEnet

FORCEnet will blend the traditional domains of operations, intelligence, and logistics and will enable adaptable and intuitive command-and-control architectures and systems to increase the speed of decisions and actions. It will provide capabilities that are fully interoperable with joint forces and will provide the Joint Force Commander flexible, adaptive options with which to address the uncertain challenges of the future.

### 2.1.2.4 The Supporting Concepts

*Sea Trial.* Sea Trial is the process of innovation designed to transition naval forces toward rapid, precise, and responsive network-centric operations through the development of concepts and technology that will deliver enhanced capabilities to the fleet. Sea Trial will deliver warfighting capabilities by experimentation, integrating concepts, technologies, and emerging information age capabilities at the fleet level. It will identify candidates with the best potential to provide the greatest enhancement to warfighting, with candidate technologies and concepts to meet fleet requirements matured through spiral development with targeted investment and rapid prototyping.

*Sea Warrior.* Sea Warrior is the Navy's commitment to the growth and development of personnel to operate the network-centric fleet of the 21st century. The process begins with an improved selection and classification of recruits and carries on through a life-long continuum of learning. Information age advancements of interactive and Web-based learning will provide for self-paced, progressive skill development and for the maintenance of skill levels in a rapidly changing environment so that all sailors are optimally trained, educated, and assigned.

*Sea Enterprise.* Sea Enterprise addresses the challenge of resourcing tomorrow's fleet through improved organizational alignment, redefined requirements, and reinvested savings to buy the capabilities needed to transform the Navy. Combining past experience with information age practices and systems, the Navy will streamline organizations and divest noncore functions so as to enhance current operations and increase the investment in future warfighting capabilities.

### 2.1.2.5 The Application

*Global Concept of Operations.* All of the elements of the Sea Power 21 vision described above contribute to what is termed a Global Concept of Operations, which responds to the complex emerging security environment "with the ability to respond to a broad range of scenarios" so as "to sustain homeland defense, provide forward deterrence in four theaters, swiftly defeat two aggressors at the same time, and deliver decisive victory in one of those theaters."[19] This concept of operations builds on FORCEnet to multiply the impact of naval forces by dispersing combat striking power into "independent operational groups capable of responding simultaneously around the world" to counter immediately any unexpected threats but capable of being "netted together for expanded warfighting effect."[20]

It is important to note that the Sea Power 21 vision is not and cannot be static. It is neither a fixed objective nor a program, however vast, that can someday be completed. Rather it is an ongoing response to a dangerous and ever-changing set of national security challenges that seeks both to employ current capabilities in new ways and to introduce innovative capabilities as quickly as possible. As Admiral Clark notes:

> [The Sea Power 21 vision] requires us to continually and aggressively reach. It is global in scope, fully joint in execution, and dedicated to transformation. It reinforces and expands concepts being pursued by other services—long range strike; global intelligence, surveillance and reconnaissance; expeditionary warfare; and light, agile ground forces—to generate maximum combat power from the joint team.[21]

## 2.2 DEFINITION OF FORCENET

### 2.2.1 One Definition, Two Elements

The FORCEnet concept clearly lies at the center of Sea Power 21 and the Marine Corps's EMW from two perspectives: that of providing the networking that the Sea Basing, Sea Strike, and Sea Shield concepts require for success and that of enabling both the Navy and the Marine Corps to deal flexibly with new challenges and to introduce new ideas and technologies into a continuing process of adaptation.

The definition of FORCEnet laid out by the CNO and a succession of SSGs and endorsed by the CMC is as follows:

> [FORCEnet is] the operational construct and architectural framework for naval warfare in the information age that integrates warriors, sensors, networks, command and control, platforms, and weapons into a networked, distributed, com-

---

[19]ADM Vern Clark, USN. 2002. "Sea Power 21 Series, Part I: Projecting Decisive Joint Capabilities," *U.S. Naval Institute Proceedings*, October, pp. 13-14.

[20]ADM Vern Clark, USN. 2002. "Sea Power 21 Series, Part I: Projecting Decisive Joint Capabilities," *U.S. Naval Institute Proceedings*, October, p. 50.

[21]ADM Vern Clark, USN. 2002. "Sea Power 21 Series, Part I: Projecting Decisive Joint Capabilities," *U.S. Naval Institute Proceedings*, October, p. 18.

bat force that is scalable across all levels of conflict from seabed to space and sea to land.[22]

This definition is a clear point of departure for all FORCEnet implementation efforts: it is already widely used, and it is broad enough to encompass the two different aspects of future FORCEnet development: (1) its role as the key enabler for Sea Power 21 and especially for the Sea Basing, Sea Strike, and Sea Shield pillars and (2) its role in defining the infrastructure of the network-centric naval forces needed to carry out the promise of Sea Power 21 and the Global Concept of Operations.

However, it is important to note that the definition points to FORCEnet as both "operational construct and architectural framework." That is, two different elements and two distinct tasks are envisioned in FORCEnet: the tasks for the one element are to define and implement the operational construct; the tasks for the other element are to design and build the architectural framework that will enable that operational construct to work, and with it, the Global Concept of Operations and the Sea Basing, Sea Strike, and Sea Shield pillars of Sea Power 21.

### 2.2.1.1 The Operational Construct

The operational construct for FORCEnet is in essence the concept of employment of FORCEnet for realizing network-centric operations and applying that concept to "naval warfare in the information age."[23] In this context, the operational construct is inseparable from the Sea Basing, Sea Strike, and Sea Shield pillars of Sea Power 21 for which it is the critical enabler. Moreover, in this same context FORCEnet both supports and is supported by the concepts contained in the Sea Warrior, Sea Trial, and Sea Enterprise initiatives.

The operational construct of FORCEnet is very much in the domain of the naval forces' operator and warfighter and is at the root of their efforts to integrate "warriors, sensors, networks, command and control, platforms and weapons"[24] into concepts of operations that optimize each so as to generate the overwhelming effects from the sea that can be "decisive in peace and in war."[25]

---

[22]VADM Richard W. Mayo, USN; and VADM John Nathman, USN. 2003. "Sea Power 21 Series, Part V: ForceNet: Turning Information into Power," *U.S. Naval Institute Proceedings,* February, p. 42.

[23]VADM Richard W. Mayo, USN; and VADM John Nathman, USN. 2003. "Sea Power 21 Series, Part V: ForceNet: Turning Information into Power," *U.S. Naval Institute Proceedings,* February, p. 42.

[24]VADM Richard W. Mayo, USN; and VADM John Nathman, USN. 2003. "Sea Power 21 Series, Part V: ForceNet: Turning Information into Power," *U.S. Naval Institute Proceedings,* February, p. 42.

[25]ADM Jay Johnson, USN. 1997. "Anytime, Anywhere," *U.S. Naval Institute Proceedings,* November, p. 50.

To this end, the operational construct for FORCEnet will need to be defined in terms of how the concepts and requirements of Sea Basing, Sea Strike, and Sea Shield are integrated into the desired FORCEnet capability. Further, both FORCEnet and these Sea Power 21 pillars will need to be defined in terms of how they apply to emerging joint concepts for, among other things, forcible entry, undersea warfare, sea basing, and effects-based operations and of how they might apply to multinational and coalition operations.

Although it seems clear that FORCEnet will play a critical role in all of the concepts and operations described above, it is evident that the requisite concepts for operational employment have just begun to be developed. Moreover, the security environment to which these concepts must respond is very different from that of the Cold War and is likely to continue mutating in response to the operational requirements that will stem from ever-changing asymmetric challenges and from more traditional threats. This volatile and dangerous security environment was the driver for both the original SSG work and for the Sea Power 21 vision. Neither the volatility nor the danger of the world, and thus the pressure for continual adaptation by naval and joint forces, is likely to diminish in coming years.

Given these strategic and operational drivers, Sea Basing, Sea Strike, and Sea Shield will require a continuous process of operational innovation in order to meet the challenges of new circumstances and new threats—sometimes on very short notice. The FORCEnet operational construct, therefore, will not be a fixed end state. Instead it will be a template that will change as operational requirements and the Sea Power 21 pillars themselves change. Accordingly, the starting operational construct must be broad enough to permit a continuing process of conceptual adaptation so as to meet the needs of operators and warfighters, and it must be flexible enough to enable this innovation to incorporate and build upon new technologies. This situation points to the need for a spiral of FORCEnet and Sea Power 21 concept development to be driven by the joint and naval operational demands sparked by changes in the security environment but enabled by changes in the technological environment, particularly in information, sensor, and weapons technologies.

### 2.2.1.2 The Architectural Framework

The architectural framework of FORCEnet describes the networking, technology, and other infrastructure needed for network-centric operations by naval and joint forces. Although there has been a tendency to think of this framework solely in terms of an information infrastructure, it encompasses two distinct parts: (1) the information infrastructure and (2) the weapons, sensor, and command architectures. This two-faceted framework will of necessity have an acquisition focus and be described in terms of architecture and infrastructure. It must also have sufficient detail to permit the creation and execution of the programs needed to create the evolving Sea Power 21 infrastructure. And, it must have sufficient flexibility to enable emerging strategic and operational needs to drive technology and program development.

The role of FORCEnet in the operational construct described above implies a set of deliverables necessary to govern and implement the operational concept spiral and to support increasingly rapid and accurate decision making. When these deliverables are mature enough to provide an operational capability of sufficient utility to the operational forces to justify the cost, the deliverables are integrated into mission and engagement packs.

The deliverables can be divided into two broad categories: (1) design and implementation requirements and (2) technology and systems to be provided:

1. Design and implementation requirements
   - Information infrastructure
     — Network architecture
     — Joint standards
     — Design reference mission
     — Common data packaging
   - Weapons, sensor, and command architectures
     — Modeling and simulation tools and experiments
     — Seamless interoperability
     — Strengthened security
     — Sea Warrior training and education
     — Fire control loops
2. Technology and systems to be provided
   - Information architecture
     — Network management tools
     — Information transport
     — Operations and maintenance training facility
   - Weapons, sensor, and command architectures
     — Networked sensors/information and knowledge
     — Analysis tools
     — Decision aids
     — Battle management systems, including a common operational picture.

The list of specific systems and technologies required for FORCEnet implementation will evolve and change as both the technology and operational concepts and, with them, FORCEnet itself, change. This idea can be broadly described as a routine and continual information exchange between the operational construct and the architectural framework.

### 2.2.2 Processes and Descriptive Items

It should be noted that the discussion in this report indicates that FORCE-net—as concepts of employment and architectures—is composed of those processes and descriptive items that guide the implementation and realization of network-centric capabilities in the force rather than being the implemented components themselves. The implemented components are referred to using "FORCEnet" as a modifying term—for example, "the FORCEnet Information Infrastructure."

## 2.3 DUAL-SPIRAL COEVOLUTION

The definition of FORCEnet with its twin focuses of operational construct and architectural framework implies that the evolution of FORCEnet cannot be driven by changes in either of these elements alone. Instead, FORCEnet must reflect a coevolution of both the operational construct and the architecture. The concept of FORCEnet coevolution points to the need to integrate two different development spirals—one centered on the Sea Basing, Sea Strike, and Sea Shield pillars of Sea Power 21 and on all of the operational or nonmateriel aspects of FORCEnet, and the other centered on the technology and architectural aspects. If FORCEnet, EMW, and Sea Power 21 are to succeed, neither of these spirals can proceed independently. Rather, each must continually support, interact with, and inform the other so that the developments in one spiral can drive the evolution of the other cycle and vice versa. New operational requirements can then stimulate and focus new technology while new technology developments enable new operational solutions—an interaction as suggested in Figure 2.1.

The innovation potential of the two spirals as well as the agility of both FORCEnet and Sea Power 21 in responding to the operational needs of a rapidly changing world security environment will derive from the interactions between the spirals. However, fostering these interactions poses a problem: each spiral, almost by definition, would be expected to proceed not only independently of the other but also at a different pace, posing a challenge akin to that of matching the spirals of a pair of still-gyrating Slinkies. This challenge suggests that it will be necessary to provide some framework in which the interactions can occur both on a regular basis and in response to specific demands.

### 2.3.1 Mission, Engagement, and Option Packs

One possible framework for interactions between the development spirals might be based on a concept provided by SSG 22: a concept that the group termed *engagement packs*, an ongoing succession of specific FORCEnet component capabilities at the individual system level (for example, information displays) that apply emerging technologies to emerging warfighter needs but are part of

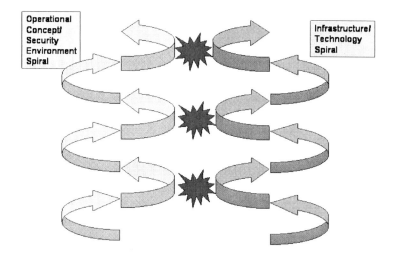

FIGURE 2.1  The concept of FORCEnet coevolution involves the integration of two development spirals: that of the operational concept and that of the infrastructure and technology.


overall FORCEnet evolution.[26] This concept might be expanded toward *mission packs*, broad sets of capabilities supporting major Sea Power 21 concepts such as Sea Basing; and toward *options packs*, which respond to urgent warfighter needs using only those military and civilian technologies immediately available off the shelf. However, a security environment of "frequent crises, often with little warning" and threats that are "varied and deadly"[27] underline the need for an additional kind of interaction: generating immediate options from existing capabilities and technologies to deal with urgent warfighter needs, that is, *capability options packs*. These interactions between the conceptual and technological spirals can occur on multiple levels, from the tactical operational concepts or specific program or system level to the level of broad conceptual development.

These considerations suggest a process in which there are three distinctly different kinds of interactions between the spirals. Each type of interaction is independent of the others but nonetheless contributing to the overall FORCEnet evolution, as illustrated in Figure 2.2.

[26]ADM James R. Hogg, USN (Ret.), Director, CNO Strategic Studies Group, personal communication, November 9, 2005.

[27]ADM Vern Clark, USN. 2002. "Sea Power 21 Series, Part I: Projecting Decisive Joint Capabilities," *U.S. Naval Institute Proceedings,* October, p. 3.

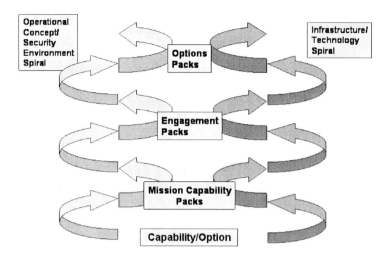

FIGURE 2.2 FORCEnet coevolution: interactions between dual development spirals.

### 2.3.1.1 Mission Packs

At the level of the Sea Basing, Sea Strike, Sea Shield, and EMW concepts and the adaptation of joint and naval strategy and doctrine to security challenges emerging over the long term, the coevolutionary interaction would come in the form of mission packs. These packs would absorb the impact of broad sets of new technologies, systems, and processes from the materiel spiral on the FORCEnet concept development; they would also encompass new requirements for FORCE-net produced by an evolving operational spiral upon which technologies are pursued and would help determine how they are prioritized in the technology spiral. The mission packs would, therefore, reflect both the evolution of the concepts of Sea Basing, Sea Strike, Sea Shield, and EMW and their impact on the FORCEnet concept, together with the interaction between this evolution and the larger set of evolving joint operations concepts and strategies. For example, the FORCEnet technology and program needs for the implementation of Sea Basing might be expected to relate in part to how Sea Shield was used to protect sea-based forces and to the ways in which Sea Basing would support Sea Shield missions beyond simply protecting those forces. This relationship in turn would depend on how the Sea Basing and Sea Shield concepts figured in joint concepts for, among other things, joint forcible-entry operations and joint and coalition sea basing. And, these relationships would in turn be shaped by the requirements of the changing security environment, such as declining access to ports, airfields, and facilities in likely conflict areas, or by changes in national policy.

The mission packs would permit these broad questions to be considered in interrelated sets. The sets would respond to the security environment, would include a broad range of naval and joint and national factors, and would both drive the technology and materiel tasking and exploit existing advances in order to deal with emerging challenges more effectively.

### 2.3.1.2 Engagement Packs

At the level of more-specific and midterm operational needs and of specific technologies and systems, the coevolutionary interaction might come in the form of engagement packs. These would cover a broad range of interactions, falling into two general types:

1. Like the mission packs, engagement packs might look at an end-to-end problem area from two perspectives: that of the interaction of various warfare areas and that of the linkage of multiple, different technologies and systems to deal with that problem. For example, a problem of air and missile defense for sea-based forces would encompass many different platforms, sensor systems, and weapons and would consider solutions from many different technologies and systems.

2. The engagement packs might also consider much more specific problems involving the application of one particular technology or system to a certain problem or capability. For example, they might consider the impact of improved information displays or decision aids on air defense.

Both of these kinds of engagement packs would appear to fit well into the current Sea Trial process, while the first might work into an expanded process that perhaps included elements of war gaming.

### 2.3.1.3 Options Packs

The object of the options packs is to provide sufficient agility in the dual-spiral development process to permit rapid adaptation of the concepts and technologies needed for meeting the ad hoc challenges and rapidly emerging threats that have become the hallmarks of the post–9/11 world security environment. The requirement to respond to emerging, urgent operational needs (e.g., for the detection and detonation of roadside bombs in Iraq) differs from the requirement to respond to needs addressed by mission and engagement packs in that the former entails finding an immediate and perhaps one-time solution, using only the assets and technologies at hand, whether from civilian industry or from government. Whereas the engagement packs fit well into the Sea Trial process, the options packs, as urgent reactions to real-world problems, would largely be proven in the field. Thus, a feedback process including lessons learned will be needed to ensure that operational-technical solutions of lasting value are captured, evaluated in terms of their potential to support long-term concept or program development, and exploited as feasible.

## 2.4  FINDINGS AND RECOMMENDATIONS

### 2.4.1  Findings

Following are the findings of the committee from its considerations related to defining FORCEnet:

• The concepts embodied in Sea Power 21 and EMW, and especially in Sea Basing, are likely to be primary drivers in defining the operational requirements of the Navy and Marine Corps for the next 20 to 30 years.

—The Sea Power 21 vision is the Navy's response to an altered world security environment, a response driven by a consistent, decade-long evolution of Navy–Marine Corps thinking that has transcended any one CNO, CMC, or administration. Since that altered security environment will continue to dictate joint and naval operational requirements for at least the next 20 to 30 years, Sea Power 21 and EMW or some close variants are likely to persist as the defining framework for naval operations. The impetus of the changed security environment is likely to be particularly strong in the case of Sea Basing because of its close link to emerging joint solutions and the requirements of national military strategy.

—FORCEnet is an integral part of the Navy–Marine Corps response to this security environment and the major enabler of both Sea Power 21 and EMW. Its development cannot be separated from the naval Services' response to the changing security environment without losing FORCEnet's coherence and relevance.

• The current definition of FORCEnet as promulgated by the CNO and successive SSGs is adequate as the point of departure for FORCEnet implementation, but further elaboration of the meaning of and distinction between the terms "operational construct" and "architectural framework" is needed.

—The current definition of FORCEnet is consistent with Sea Power 21 and EMW and with the direction of naval thought over the past decade, and it is broad enough to permit continued concept development in areas such as Sea Basing. It also offers room for amplification, for example, in the more precise definition of a network-centric architectural framework for the naval Services.

—The current definition is in such wide use that the introduction of any new definition would likely be confusing and potentially counterproductive.

• The set of requirements toward which FORCEnet must build is not and cannot be static. FORCEnet will never be completed. Rather, in its full scope FORCEnet could be considered to impact almost all aspects of the naval Services. Thus, FORCEnet will need to be able to offer an ongoing response to new operational requirements and technological possibilities that will change even as FORCEnet is implemented.

—The security environment that both naval and joint concepts of operations address will continue to change, as will the threats and demands upon naval and joint forces.

—These changes will drive the further evolution of Sea Power 21 and EMW. Accordingly, Sea Basing, Sea Strike, Sea Shield, and FORCEnet are still evolving and must continue to evolve, as must the emerging joint concepts of operations.

—The technology base and infrastructure upon which the FnII will draw are likewise evolving as the ongoing, multifaceted technology revolution centered in civilian industry continues apace. The potential solution space for meeting requirements will, therefore, evolve as technologies and systems improve.

—The FORCEnet challenge is not to build toward a distant, fixed requirement, but to adjust to the inevitable changes in real-world requirements and to tap new technologies as they emerge.

### 2.4.2 Recommendations

Based on the finding presented above and on the issues described in this chapter, the committee recommends the following.

• **Recommendation** for the Office of the Chief of Naval Operations (OPNAV), the Naval Network Warfare Command (NETWARCOM), and the Marine Corps Combat Development Command (MCCDC): Articulate better the meaning of the terms "operational construct" and "architectural framework" in the description of FORCEnet and indicate how FORCEnet implementation measures relate to each of these concepts.

• **Recommendation** for OPNAV, NETWARCOM, and MCCDC: Make clear that FORCEnet applies to the entire naval force and not just to its information infrastructure component. In so doing, the organizations should specifically indicate that the concepts of employment and the architectures developed must apply to the operation of the whole force and not just to its information infrastructure component.

• **Recommendation** for DON: Develop, maintain, and institutionalize FORCEnet coevolution along the development spirals of both the operational construct and the architectural framework.

—The development spirals of both the operational construct and the architectural framework need to interact regularly with one another if FORCEnet is to succeed.

—Mission, engagement, and options packs should be used to provide mechanisms for ensuring and exploiting interaction between the two development spirals.

# 3

# Joint Capability Development and Department of Defense Network-Centric Plans and Initiatives

## 3.1 INTRODUCTION

> Use of FORCEnet . . . facilitates integrated Naval Forces and operations that are fully interoperable with the other joint forces. It will focus on creating information networks with new levels of connectivity and integration, which will integrate the force into a joint information network.[1]

The leadership of the Navy and Marine Corps is committed to providing "flexible, persistent, and decisive warfighting capabilities as part of a joint force."[2] At the same time, leadership in OSD is pressing the military departments and the Services to become more responsive to OSD's and combatant commanders' priorities for developing warfighting capabilities and implementing network-centric operations and processes.[3] This urging has resulted in the OSD and the Joint Staff's changing, during 2003, all higher-level guidance relevant to FORCEnet implementation. The combination of the naval leadership's commitment to joint warfighting, which includes coalition operations with allies and security partners, and the broader DOD leadership's commitment to strengthening jointness and

---

[1]ADM Vern Clark, USN, Chief of Naval Operations; and Gen Michael W. Hagee, USMC, Commandant of the Marine Corps. 2003. *Naval Operating Concept for Joint Operation*, Department of the Navy, Washington, D.C., September 22, p. 6.

[2]ADM Vern Clark, USN, Chief of Naval Operations; and Gen Michael W. Hagee, USMC, Commandant of the Marine Corps. 2003. *Naval Operating Concept for Joint Operation*, Department of the Navy, Washington, D.C., September 22, cover letter.

[3]Office of the Secretary of Defense. 2003. Memorandum: "Legislative Priorities for FY 2005," U.S. Department of Defense, Washington, D.C., September 24.

developing processes more responsive to combatant commanders' needs, affects all aspects of FORCEnet implementation: namely, requirements prioritization and acquisition, concept development and experimentation, testing, and training. For FORCEnet to fulfill its intended function, it also must integrate into the broader DOD information infrastructure represented by the GIG. This chapter addresses the changes in broader DOD joint capability development and network-centric plans and initiatives that have direct implications for a FORCEnet implementation strategy.

## 3.2 REQUIREMENTS PRIORITIZATION AND ACQUISITION

### 3.2.1 Defense Planning Process

The Secretary of Defense (SECDEF) has implemented a new Defense Planning Process to provide the highest-level guidance within the DOD on resource prioritization and programming across the department.

In March 2003, the SECDEF commissioned a study led by former Undersecretary of Defense for Acquisition, Technology, and Logistics, E.C. "Pete" Aldridge, Jr., to provide streamlined processes, alternative functions, and organizations to better integrate defense capabilities in support of joint warfighting objectives.[4] The study found that:

• Services dominate the current requirements process, focusing on Service programs and platforms rather than on the capabilities required to accomplish combatant command missions; this situation results in an inaccurate picture of joint needs and an inconsistent view of priorities and acceptable risks across the DOD.

• Service planning does not consider the full range of solutions available to meet joint warfighting needs; alternative ways to provide equivalent capabilities receive inadequate attention, particularly if the alternative solutions reside in a different Service or defense agency.

• The resourcing function focuses the efforts of senior leadership on fixing problems at the end of the process, rather than its being involved early in the planning process.

• OSD programming guidance exceeds available resources and does not provide realistic priorities for joint needs, resulting in a program that does not best meet joint needs or provide the best value for the nation's defense investment.

---

[4]Joint Defense Capabilities Study Team. 2004. *Joint Defense Capabilities Study: Improving DOD Strategic Planning, Resourcing, and Execution to Satisfy Joint Requirements*, Final Report, January, Department of Defense, Washington, D.C.

In October 2003, the SECDEF issued a memorandum directing the establishment of a Joint Capabilities Development Process to examine, on the basis of the Aldridge study, major issues in support of the development of the 2006–2011 programs and budget.[5] The new process differs from the current process in the following respects:

• Combatant commanders are assigned a much larger role in shaping the defense strategy articulated in the Strategic Planning Guidance (SPG), which replaces the Defense Planning Guidance and focuses on strategic objectives, priorities, and risk tolerance, rather than on programmatic solutions. The SPG initiates the planning process and dictates those areas in which joint planning efforts must focus.

• An Enhanced Planning Process (EPP) supports the assessment of capabilities for meeting joint needs; these are identified primarily through combatant command operational plans and operating concepts. The Services and the OSD retain responsibility for identifying nonwarfighting needs.

• A forthcoming document, Joint Programming Guidance, will reflect decisions made in the EPP and provide fiscally executable guidance for the development of programs, with the intent of simplifying the remainder of the resourcing process and reducing the scope and level of effort required for program and budget reviews.

The new Defense Planning Process is immature, and the initial results will not be evident until the fall of 2004 (some months after this writing). Issuing the SPG took longer than anticipated, and the EPP is ongoing. However, the intent of the SECDEF to produce fiscally feasible Joint Programming Guidance directing the Services to acquire specific capabilities is clear.

### 3.2.2 Joint Capabilities Integration and Development System

The new Joint Capabilities Integration and Development System (JCIDS)[6] replaces the former requirements-generation system with a process that emphasizes joint concepts and, using those concepts, capabilities-based planning. The Joint Staff led the development of the JCIDS approach, and it preceded the Aldridge study, but its motivations are similar to those of the study. As its name implies, the new process is intended to provide substantive improvements in interoperability among components of joint forces in future battles. Figure 3.1 depicts the JCIDS process as it was envisioned.

---

[5]Office of the Secretary of Defense. 2003. Memorandum: "Initiation of a Joint Defense Capabilities Process," U.S. Department of Defense, Washington, D.C., October 31.

[6]Chairman of the Joint Chiefs of Staff. 2003. Chairman of the Joint Chiefs of Staff Instruction, CJCSI 3170.01C, "Joint Capabilities Integration and Development System," The Pentagon, June 24.

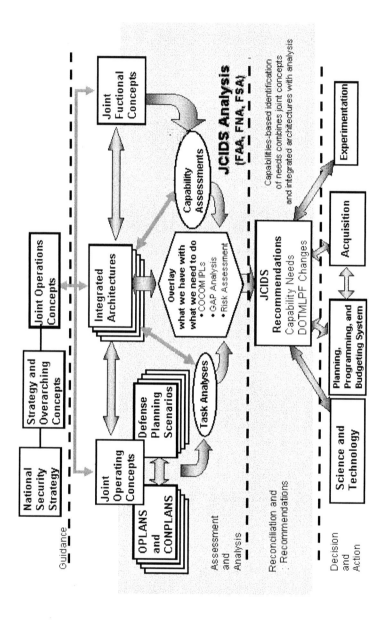

FIGURE 3.1 The Joint Capabilities Integration and Development System's (JCIDS's) top-down process for identifying capability needs. NOTES: OPLANS, operational plans; CONPLANS, concept plans; COCOM, combatant command; IPL, integrated priority list; FAA, functional area analysis; FNA, functional needs analysis; FSA, functional solutions analysis; DOTMLPF, doctrine, organization, training, materiel, leadership and education, personnel and facilities. SOURCE: CJCSI 3170.01C.

The JCIDS process alters the former requirements-generation system in order to emphasize capability needs rather than threat responses. It replaces Mission Needs Statements with Initial Capabilities Documents, and Operational Requirements Documents are replaced with Capability Development Documents and Capability Production Documents.

The JCIDS process applies to all acquisition categories (ACATs). All of the new capabilities documents are to be sent to a "gatekeeper" on the Joint Staff to determine whether they fit into various categories—being of interest to the Joint Requirements Oversight Council (JROC)[7] (all ACAT 1/1A programs),[8] having joint impact, requiring joint integration, or being independent—depending on the extent to which the gatekeeper deems the program to impact joint concepts and operations. Only those programs categorized as "independent" require no further joint certification. Functional capabilities boards are to analyze gaps and seams among those programs that have joint impact or require joint integration and therefore require joint certification. Though FORCEnet is not a program, FORCEnet-related programs will have joint impact and therefore will entail joint certification. This status will link FORCEnet development even more closely to DOD efforts to create joint battle management command and control (JBMC2) and the GIG.

### 3.2.3 Joint Battle Management Command and Control

The Deputy Secretary of Defense's Management Initiative Decision 912 in early 2003 directed JFCOM to lead efforts to strengthen the organizing, training, and equipping of joint battle management command and control capabilities for combatant commanders.[9] JBMC2 is deemed to consist of the processes, architectures, systems, standards, and command-and-control operational concepts employed by the Joint Force Commander. The JBMC2 effort is governed by a board of directors consisting of flag officers and chaired by JFCOM, with representatives from the combatant commands and Services in the core group, and a wider

---

[7]Under the current process, requirements are matched against specific military needs at the Pentagon as the DOD develops its share of the president's annual budget request. These requirements are vetted by the Pentagon's Joint Requirements Oversight Council, a body that includes the vice chief of staff of each of the uniformed Services. The council has the power to approve or defer requirements.

[8]The ACAT designations (I, II, III, etc.) are established for all the military Services by DOD Instructions 5000.1 and 5000.2 and their Service-specific supplements. For additional information, see http://www.acquisition.navy.mil. Accessed July 24, 2004.

[9]Deputy Secretary of Defense. 2003. Management Initiative Decision 912: "Charter for Joint Battle Management Command Control (JBMC2) Board of Directors (BOD)," U.S. Department of Defense, Washington, D.C., January 7.

group of advisers having responsibilities for affected programs. A JBMC2 road-map is in preparation. The goal of the roadmap is as follows:

> [to] develop a coherent and executable plan that will lead to integrated JBMC2 capabilities and interoperable JBMC2 systems that in turn will provide net-worked joint forces:
>
>> • Real-time shared situational awareness at the tactical level and common shared situational awareness at the operational level;
>> • Fused, precise, and actionable intelligence;
>> • Decision superiority enabling more agile, more lethal, and survivable joint operations;
>> • Responsive and precise targeting information for integrated real-time of-fensive and defensive fires; and
>> • The ability to conduct coherent distributed and dispersed operations, in-cluding forced entry into anti-access or area-denial environments.
>
> This roadmap will be the vehicle for prioritizing, aligning and synchronizing Service JBMC2 architectural and acquisition efforts.[10]

The draft JBMC2 roadmap calls for plans to be complete by the beginning of 2006, followed promptly by "cluster" tests, presumably tied to joint mission threads such as joint close-air support. The roadmap addresses FORCEnet in the context of providing a single integrated maritime picture, with ashore network integration accomplished in 2008 and afloat network integration accomplished in 2010. However, the draft roadmap has only a placeholder for the detailed descrip-tion of FORCEnet.[11] From the description of FORCEnet in the draft roadmap, it is not clear what influence the JBMC2 board of directors will attempt to exert on the development of the FnII.

### 3.2.4 Joint Lessons Learned from Recent and Ongoing Operations

On the basis of lessons learned from operations and experimentation in determining defense priorities for capabilities development and programming, the new Defense Planning Process and JCIDS identify a greater role for combat-ant commands' integrated priority lists. More directly, the global war on terror-ism and the lessons and clear shortfalls evident in operations in Afghanistan and Iraq are becoming major drivers both for resource-allocation priorities and for the creation of processes that can respond faster to operational needs.

---

[10]Under Secretary of Defense for Acquisition, Technology, and Logistics, and U.S. Joint Forces Command. 2004. *Joint Battle Management and Command and Control Roadmap*, Version 2.0, U.S. Department of Defense, Washington, D.C., February 27, p. xv.

[11]Under Secretary of Defense for Acquisition, Technology, and Logistics, and U.S. Joint Forces Command. 2004. *Joint Battle Management Command and Control Roadmap*, Version 1.0, U.S. Department of Defense, Washington, D.C., May.

In conjunction with OIF, JFCOM positioned teams of analysts at the major joint command headquarters for the express purpose of gathering joint operational insights on a comprehensive scale as the operations unfolded, rather than collecting impressions following the operations. Based on these observations the commander of JFCOM testified:

> The fundamental point is that our traditional military planning and perhaps our entire approach to warfare have shifted. The main change, from our perspective, is that we are moving away from employing Service-centric forces that must be de-conflicted on the battlefield to achieve victories of attrition to a well-trained, integrated joint force that can enter the battlespace quickly and conduct decisive operations with both operational and strategic effects. Joint Force Commanders today tell me that they don't care where a capability comes from so long as it meets their warfighting needs. They also tell me that "it's not the plan, it's the planning." They understand that the ability to plan and adapt to changing circumstances and fleeting opportunities is the key to rapid victory in the modern battlespace.[12]

The Chairman of the Joint Chiefs of Staff (CJCS) has directed the commander of JFCOM to carry out the following:

- Aggregate key joint operational and interoperability lessons reported by combatant commands, defense agencies, and the Services during OIF and the war on terrorism;
- Analyze, categorize, and prioritize these lessons, working with functional capabilities boards; and
- Convey recommendations of materiel and nonmateriel approaches for remedies to shortfalls indicated by the lessons learned to the JROC as the basis for recommendations to the SECDEF.[13]

This effort is leading to the establishment of a permanent JFCOM organization on lessons learned (the Joint Center for Operations Analysis and Lessons Learned) linked to its requirements division (J8), which also, chairs the JBMC2 board of directors. This organization will document joint lessons-learned efforts, produce proposals regarding program changes, and have designated agents track their implementation.[14] The intention is to provide shortfall remedies to deploying forces rather than to have them learn similar lessons.

---

[12]ADM Edmund P. Giambastiani, Jr., Commander, U.S. Joint Forces Command, and Supreme Allied Commander, NATO. 2003. Statement on Transformation before the 108th U.S. Congress, House Armed Services Committee, October 2.

[13]Chairman of the Joint Chiefs of Staff. 2003. Memorandum for the Commander, U.S. Joint Forces Command, CM-1318-03, "Expansion of Joint Lessons Learned—The Next Step," The Pentagon, October 31.

[14]Commander, U.S. Joint Forces Command. 2003. Memorandum, "Expanding the Lessons Learned Effort," U.S. Department of Defense, December 9.

### 3.3 JOINT CONCEPT DEVELOPMENT AND EXPERIMENTATION

To ensure joint forces are truly interoperable and complementary in the future, the Sea Services will be fully engaged in Joint Concept Development and Experimentation (JCD&E).[15]

The Navy and Marine Corps have been coordinating their concept development and experimentation (CD&E) activities more closely over the past 5 years than they did before.[16] Beginning in 2003, with the JFCOM Pinnacle Impact 03 and Unified Quest 03 experiments, they have increased their respective efforts to participate in JCD&E activities and in efforts to "showcase the utility of Naval concepts during other Services' Title X war games."[17]

#### 3.3.1 Joint Concept Development

The JCIDS process calls for programs to be organized around the capabilities needed to execute joint concepts. The process (depicted in Figure 3.1) derives guidance from the National Security Strategy and amplifying documents. Ideally, from these sources come an overarching Joint Operating Concept and subordinate joint operating and functional concepts: these concepts would inform the development of integrated architectures as a basis for analysis and risk assessment in determining capability needs, and they would inform the resourcing and joint experimentation needed for assessment.

In November 2003, the SECDEF issued the "Joint Operations Concept," (JOpsC) as the overarching document that contains the following:[18]

• A description of how the Joint Force intends to operate within the next 15 to 20 years;

• The conceptual framework to guide future joint operations and joint, Service, combatant command, and combat support defense agency concept development and experimentation; and

• The foundation for the development and acquisition of new capabilities through changes in doctrine, organization, training, materiel, leadership and education, personnel and facilities (DOTMLPF).

---

[15]ADM Vern Clark, USN, Chief of Naval Operations; and Gen Michael W. Hagee, USMC, Commandant of the Marine Corps. 2003. *Naval Operating Concept for Joint Operation*, U.S. Department of Defense, Washington, D.C., September 22, p. 20.

[16]ADM Vern Clark, USN, Chief of Naval Operations; and Gen Michael W. Hagee, USMC, Commandant of the Marine Corps. 2003. *Naval Operating Concept for Joint Operation*, U.S. Department of Defense, Washington, D.C., September 22, p. 20.

[17]ADM Vern Clark, USN, Chief of Naval Operations; and Gen Michael W. Hagee, USMC, Commandant of the Marine Corps. 2003. *Naval Operating Concept for Joint Operation*, U.S. Department of Defense, Washington, D.C., September 22, p. 20.

[18]Office of the Secretary of Defense. 2003. Memorandum, "Joint Operating Concept," U.S. Department of Defense, November.

The JOpsC includes a taxonomy of subordinate concepts, including joint operating concepts (JOCs), joint functional concepts, and enabling concepts (now termed joint integrating concepts). In this construct of concepts, "there is no hierarchy to operating, functional or enabling concepts—they must all inform and interrelate with each other."[19] The "Joint Operating Concept" describes the function of JOCs as follows: "focusing at the operational-level, JOCs integrate functional and enabling concepts to describe how a JFC [Joint Force Commander] will plan, prepare, deploy, employ and sustain a joint force given a specific operation or combination of operations."[20] The purpose of functional concepts, having the JOpsC and JOCs for their operational context, is to amplify a particular military function and apply it broadly across the range of military operations.

The DOD's "Transformation Planning Guidance," published in April 2003, had directed that "the CJCS, in coordination with Commander, JFCOM, will initially develop one overarching joint concept and direct the development of four subordinate JOCs: homeland security, stability operations, strategic deterrence, and major combat operations."[21] The JOpsC described the scope of these concepts and identified four initial functional concept categories of joint command and control: battlespace awareness, force application, focused logistics, and protection. A functional capabilities board for each of these categories (and a fifth recently added network-centric warfare category) serves both to articulate the functional concept and to certify programs deemed to have joint impact or to require joint interoperability. The Naval Operating Concept for Joint Operations aligns its concepts of Sea Strike, Sea Shield, Sea Basing, and FORCEnet with the Joint Vision 2020[22] concepts of precision engagement, dominant maneuver, full dimensional protection, focused logistics, and joint command, control, communications, computers, intelligence, surveillance, and reconnaissance (C4ISR), from which the new functional capabilities categories were derived.

Combatant commands have been assigned the lead in writing the operating concepts (JFCOM: major combat operations, stability operations, joint forcible-entry operations; U.S. Strategic Command (STRATCOM): strategic deterrence; NORTHCOM: homeland security). The subordinate concept documents are in preparation and review, awaiting JROC approval. The Navy and Marine Corps are participating in the development of these concepts to ensure congruence between these concepts and the Naval Operating Concept for Joint Operations.

---

[19]Office of the Secretary of Defense. 2003. Memorandum, "Joint Operating Concept," U.S. Department of Defense, November, p. 18.

[20]Office of the Secretary of Defense. 2003. Memorandum, "Joint Operating Concept," U.S. Department of Defense, November, p. 18.

[21]Office of the Secretary of Defense. 2003. "Transformation Planning Guidance," U.S. Department of Defense, April, p. 15.

[22]GEN Henry H. Shelton, USA, Chairman of the Joint Chiefs of Staff. 2000. *Joint Vision 2020*, The Pentagon, Washington, D.C., June, p. 13.

In its initial efforts to implement the JCIDS process, the Joint Staff found that it needed to establish a set of joint integrating concepts to derive integrated architectures from the operational and functional concepts. Figure 3.2 illustrates the latest approach for shifting from a bottom-up approach driven principally by the Services to a top-down approach driven principally by the OSD leadership and combatant commanders. Noteworthy is that the chart indicates that both the bottom-up and top-down approaches coexist today, but that the top-down approach is intended to drive the Service transformation roadmaps and concepts in the future. Also notable is that settling upon a set of joint integrating concepts has proven a challenge. Two concepts selected for early development are those of undersea superiority (for which the Pacific Command (PACOM) would have the lead in writing the operational concepts) and sea basing (for which the U.S. Transportation Command would have the lead).

### 3.3.2 Joint Experimentation

The DOD's Joint Experimentation Program began in May 1998 when the SECDEF designated the U.S. Atlantic Command (which became JFCOM in

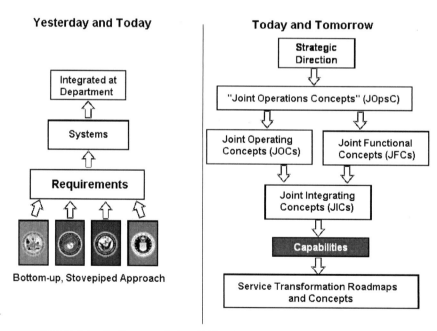

FIGURE 3.2 Joint Staff approach for shifting from a bottom-up approach driven principally by the Services to a top-down capabilities-based methodology. SOURCE: U.S. Joint Chiefs of Staff.

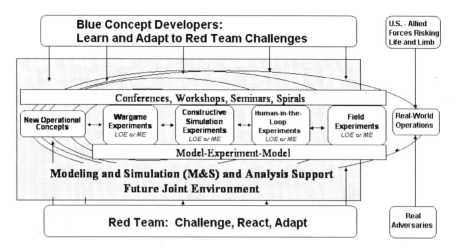

FIGURE 3.3 DOD's concept of a continuous process of interactive experimentation for the joint experimentation process. NOTES: LOE, limited objective experiment; ME, mission execution. SOURCE: U.S. Joint Forces Command, Norfolk, Va.

October 1999) as executive agent for joint warfighting experimentation.[23] In October 1998, the U.S. Atlantic Command established a Joint Experimentation Directorate (J9) to implement this responsibility. The intent was to create a continuous process of interactive experimentation, using methods depicted in Figure 3.3. The conception was to have a continuous process of concept exploration and development, beginning with papers and assessments by subject-matter experts, moving into more rigorous war gaming, then to detailed human-in-the-loop simulations, leading to field exercises and evaluation of concepts in actual operations, with subsequent events informing the previous ones.

Red teams were recognized as an essential feature of this activity. The OSD established a Defense Adaptive Red Team as part of the Joint Warfighting Program element retained by the Deputy Under Secretary of Defense for Advanced Systems and Concepts when Joint Experimentation Program funds were transferred to JFCOM. This team employs a wide variety of regional, country, and local-area expertise, in addition to selected retired military officers, to play the role of nontraditional adversaries in various DOD-sponsored games. JFCOM also established a red team as part of its J9.

[23]U.S. Department of Defense. 1998. News Release 252-98: "U.S. Atlantic Command Designated Executive Agent for Joint Warfighting Experimentation," May 21.

The value of red teams increases with the rigor of the experiments; their use is essential for ensuring that wishful thinking does not color learning. Red teaming in general discussions is of limited use, however. The most ambitious use of red teams has occurred in the "opposition forces" at the Army's National Training Center at Fort Irwin, California, which is becoming the hub of the Joint National Training Capability. To prepare forces for operations in Afghanistan and Iraq, the scenarios have changed to include forces dealing with stability and reconstruction operations, learning to identify village leaders, assessing and assisting in reconstruction with funding and other assistance, and identifying and defeating various classes of adversaries ranging from national resistance to jihadist elements. American-Iraqis play the roles of innocents and various classes of adversaries.

The conception of the experimentation process depicted in Figure 3.3 has yet to materialize. In striving to realize its vision, JFCOM has increasingly sought Service support for its efforts.

The initial focus of the Joint Experimentation Program was at the joint operational level, on future (7 to 15)-year joint concepts.[24] The output of JFCOM experimentation was intended to be DOTMLPF recommendations to the JROC, as depicted in Figure 3.4. Upon JROC approval, these recommendations were to affect the development of joint capabilities, including Service programs.

In November 2001, the CJCS directed JFCOM to develop a Joint Experimentation Campaign Plan focused "on the development of a standing joint force headquarters model no later than the end of Fiscal Year 2004 and capable of implementation by all regional Commanders-in-Chief (CINCs) (regional combatant commanders) during FY05."[25] This directive resulted in essentially all of JFCOM's joint concept and development experimentation effort being concentrated on the Standing Joint Force Headquarters (SJFHQ).

In November 2002, the CJCS revised his joint experimentation guidance:

> The plan must incorporate a decentralized process to explore and advance emerging joint operational concepts, proposed operational architectures, experimentation and exercise activities currently being conducted by the Joint Warfighting Capabilities Assessment Strategic Topic Task Forces [run by the Joint Staff with Service participation], the combatant commands, the Services and Defense Agencies.[26]

---

[24]Rick Kass, Chief Analysis Division, U.S. Joint Forces Command, "Understanding Joint Warfighting Experimentation Methods," presentation to the Committee on the Role of Experimentation in Building Future Naval Forces, May 1, 2002.

[25]Chairman of the Joint Chiefs of Staff. 2001. Memorandum, CM-56-01, "Guidance for USCINCJFCOM Joint Experimentation," The Pentagon, November 2.

[26]Chairman of the Joint Chiefs of Staff. 2002. Memorandum, CM-635-02, "Guidance for USJFCOM Joint Experimentation," The Pentagon, November 26.

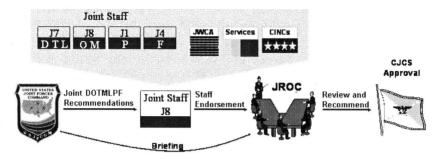

- USJFCOM submits DOTMLPF package to Joint Staff J8.
- Joint Staff J8 serves as "on-ramp" for coordination and review by Joint Staff.
- DOTMLPF package is forwarded to Joint Requirements Oversight Council.
- JROC reviews and recommends DOTMLPF package to CJCS.
- Upon approval, CJCS assigns implementation.

FIGURE 3.4 The review and approval process for joint DOTMLPF. NOTES: J7, Director, Operations Plans and Interoperability; J8, Director, Force Structure, Resources, and Assessment; J1, Director, Manpower and Personnel; J4, Director, Logistics. SOURCE: U.S. Joint Forces Command, Norfolk, Va.

This guidance kept the SJFHQ as the highest priority, but it also directed coordination with combatant commands, Services, Joint Staff, and defense agencies, as well as the inclusion of the following:

- Lessons learned from the war on terrorism;
- Joint operations in an uncertain environment and complex terrain;
- Fast-deploying joint command-and-control structures;
- Concepts to provide warfighters at all levels with improved battlespace awareness, correlation and dissemination of mission-specific information, and more closely integrated intelligence, surveillance, and reconnaissance efforts and products;
- Joint capabilities enabling the near-simultaneous, integrated, and synergistic employment and deployment of air, land, sea, cyberspace, and space warfighting capabilities, capitalizing on Service concepts and capabilities that enable forward joint forces and those based in the continental United States to deploy, employ, sustain, and redeploy in austere regions and antiaccess and area-denial environments.
- Transformational concepts of the Nuclear Posture Review (involving global strike with conventional and special forces in addition to nuclear strike, as well as global information operations); and

---

**BOX 3.1**
**Joint Forces Command Experimental Focus for**
**Fiscal Years 2004 and 2005**

| Achieving Decision Superiority | Creating Coherent Effects | Conducting and Supporting Distributed Operations |
|---|---|---|
| 1. **Achieving information superiority (anticipatory understanding)**<br>2. Decision making in a collaborative information environment<br>3. **Coalition and interagency information sharing**<br>4. Global integration<br>5. **Joint intelligence, surveillance, and reconnaissance** | 1. Information operations and information assurance<br>2. **Joint maneuver and strike:**<br>  **a. Global**<br>  **b. Operational**<br>  **c. Tactical**<br>3. **Interagency operations**<br>4. **Multinational operations**<br>5. Precise effects<br>6. **Urban operations**<br>7. Deny sanctuary<br>8. Transition/stability operations<br>9. Coercive operations | 1. **Force projection: Deployment, employment, and sustainability**<br>2. Force protection and base protection<br>3. **Counter antiaccess and area-denial (includes forcible entry operations)**<br>4. Low-density, high-demand assets<br>5. Proper decentralization |

NOTE: Emphasis in original. (The boldface items indicate priorities to be addressed in FY 2004 and FY 2005.) SOURCE: U.S. Joint Forces Command, Norfolk, Va.

---

• Current efforts to promote and develop regional component-commander-sponsored joint and multinational partnerships involving experimentation and capability-based modeling and simulation.

To implement this guidance, JFCOM J9 polled the combatant commands to obtain a sense of their priorities and coordinated with the Services to find more opportunities to explore concepts in Service war games. Polling the combatant commanders produced 308 items, which were aggregated into those illustrated in Box 3.1.[27] The highlighted items indicate priorities that were to be addressed in FY 2004 and FY 2005.

---

[27]CAPT Paul Smith, USN, Joint Concept Development and Experimentation Office, U.S. Joint Forces Command, "Naval Studies Board UC04 [Unified Course 04] Indoc," presentation to the committee on November 18, 2003.

The JFCOM J9 intent was to use Service war games based upon a common set of joint scenarios to address the priority issues. Figure 3.5 illustrates the proposed plan for FY 2004.

The OSD and Commander of JFCOM tasked the Defense Science Board (DSB) to review the Joint Experimentation Program. In its Phase I report of September 2003, the DSB found: "Of particular importance is the need to reorient JFCOM's experimentation focus from inward (largely about inviting others to participate in JFCOM events) to external (largely about participating in and influencing Service and combatant command experiments and related activities)."[28] Figure 3.5 illustrates JFCOM efforts to work with Services and other combatant commands. However, JFCOM's approach has remained one of attempting to align Service and combatant command efforts to its own agenda as opposed to supporting other organizations in experimenting with a broader set of challenges. Also, JFCOM has had difficulty realigning the resources that it devoted to the SJFHQ. While more war-gaming effort has gone into examining a broader set of warfighting concepts, the bulk of JFCOM J9's expenditures are related to the 2001 tasking to field SJFHQs to combatant commanders during FY 2005.

Aligning Sea Trial with JFCOM's concept development and experimentation could easily engulf the Navy and Marine Corps CD&E efforts. JFCOM J9 has more than 400 people on its staff. The Navy Warfare Development Command (NWDC) has about 20 people with responsibilities similar to those of the J9 staff. While supporting JFCOM with a full-time liaison officer, the MCCDC is in a situation similar to that of the Navy.

### 3.4 JOINT TESTING— THE JOINT DISTRIBUTED ENGINEERING PLANT

System testing prior to deployment will be part of any FORCEnet implementation strategy. The joint battle management command and control roadmap calls for expansion of the Joint Distributed Engineering Plant (JDEP). JDEP is a DOD-wide facility that offers Service and joint engineering, integration, and test resources to provide system-of-systems environments that are battlefield-representative, in support of developer, tester, and warfighter requirements.[29] Based on the Navy's Distributed Engineering Plant (DEP) model, JDEP is a response to operational demands for sharing data and information among systems of systems assembled to conduct military operations. The JDEP vision is to improve the

---

[28]Office of the Under Secretary of Defense for Acquisition, Technology, and Logistics. 2003. *Phase I Report of the Defense Science Board Task Force on Joint Experimentation*, U.S. Government Printing Office, Washington, D.C., September.

[29]George Rumford, Joint Distributed Engineering Plant Program Office, "Joint Distributed Engineering Plant," presentation to the committee on January 7, 2004.

62

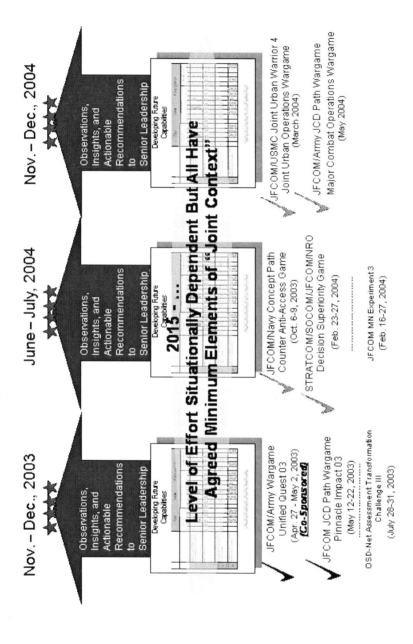

FIGURE 3.5 The Joint Concept Development war gaming plan as proposed for Fiscal Year 2004. (Check marks indicate completed actions.)
SOURCE: U.S. Joint Forces Command, Norfolk, Va.

interoperability of weapons systems, platforms, and command and control within and across the Services by providing the capability to create environments to support engineering, development, integration, testing, evaluation, and certification in a replicated battlefield environment, leveraging Service and joint combat system engineering and test sites. JDEP's goals are as follows:

- To create a capability to integrate DOD and industry laboratories, test sites, and facilities to address and resolve weapons system interoperability issues for warfighters, system developers, and testers; and
- To support users in selecting, accessing, and integrating simulations and range systems, using supporting tools, under a set of common standards and procedures.

Although not a new concept, conducting events on a joint level to synchronize efforts, resources, and assets across the Services by critical mission areas is a new approach. The JDEP technical framework comprises the components of a JDEP configuration, the interfaces, and the guidance on how to configure and apply the components to meet user needs. This technical framework is critical to cost-effectively federating simulation, hardware- and software-in-the-loop, and systems across multiple communities. Because industry is a key participant in these activities, the JDEP technical framework, including the high-level architecture (HLA), is based on industry standards, and it is being implemented using standards-based commercial products wherever possible.[30] Funds of $182 million for FY 2002–2007 are programmed for the JDEP.

Established in 2000, the JDEP program has principally supported air and missile defense events to date, but it is extending into other mission areas by working directly with acquisition programs as well as with functional capabilities boards. The OSD requires all new programs—for example, the aircraft carrier (CVN 21), the future destroyer, and the Joint Strike Fighter—to use the JDEP in order to obtain milestone B program approval.[31]

JDEP is encountering the challenge of finding exactly what technical, data, and application standard affect all current networking efforts.

---

[30]Judith Dahmann, scientific advisor to director of interoperability, Office of the Secretary of Defense for Acquisition, Technology, and Logistics and principal senior staff, MITRE Corporation; and Richard Clarke, JDEP Technical Director, Joint Interoperability Test Command, "Joint Distributed Engineering Plant Technical Framework: Applying Industry Standards to System-of-System Federations for Interoperability," undated. This paper describes the framework, the migration strategy, and progress to date in assessing and applying the technical framework.

[31]The term "milestone B approval" means a decision to enter into system development and demonstration pursuant to guidance prescribed by the Secretary of Defense for the management of Department of Defense acquisition programs.

## 3.5 JOINT TRAINING TRANSFORMATION

The Joint Training Transformation presents an opportunity for FORCEnet implementation. The objectives of the training transformation are as follows:[32]

•  To strengthen joint operations by preparing forces for new warfighting concepts,
•  To improve joint force readiness continuously by aligning joint education and training capabilities and resources with combatant command needs,
•  To develop individuals and organizations that intuitively think jointly,
•  To develop individuals and organizations that improvise and adapt to emerging crises, and
•  To achieve unity of effort from a diversity of means (drawn from Active and Reserve components of the Services and from federal agencies, international coalitions, international organizations, and state, local, and nongovernmental organizations).

Three capabilities form the foundation for the training transformation:

1. *The joint knowledge development and distribution capability* will develop and distribute joint knowledge via a dynamic, global, knowledge network that provides immediate access to joint education and training resources.
2. *The joint national training capability (JNTC)* will prepare forces by providing command staffs and units with an integrated live, virtual, and constructive training environment that includes appropriate joint context and that allows global training and mission rehearsal in support of specific operational needs. The thrusts for the JNTC are these:
•  *Improved horizontal training*—building on existing Service interoperability training,
•  *Improved vertical training*—linking component and joint command and staff planning and execution,
•  *Integration training*—enhancing existing joint exercises to address interoperability training in a joint context, and
•  *Functional training*—providing a dedicated joint training environment for functional warfighting and complex joint tasks.
3. *The joint assessment and enabling capability* will assist leaders in assessing the value of transformational initiatives with respect to individuals, organiza-

---

[32]Office of the Undersecretary of Defense for Personnel and Readiness, Director for Readiness and Training Policy and Programs. 2003. "Department of Defense Training Transformation Implementation Plan," U.S. Department of Defense, Washington, D.C., June 10.

tions, and processes by evaluating the level of joint force readiness to meet validated combatant commander requirements.

The Web-based networks that the training transformation is expected to provide will be of potential value to FORCEnet implementation. The schedule for these Web-based networks includes the following milestones for training transformation:

- Joint knowledge development and distribution capability milestones include:
    — An initial Web-based curriculum for joint military leader development by January 2004,
    — An initial Web-based delivery capability for joint individual education and training resources by February 2005, and
    — The transitioning of initial joint education and training prototype efforts to implementing organizations by March 2006, and to international coalition partners, international organizations, and nongovernmental organizations by October 2009.
- JNTC milestones include:
    — Provision of a joint context with command, control, communications, computers, intelligence, surveillance, and reconnaissance to major Service training events and joint command and staff training events by October 2005;
    — Use of the joint training system to link lessons learned from military operations, joint training, experimentation, and testing to the development and assessment of joint operational capabilities by October 2005;
    — Demonstration of a deployable JNTC and mission-rehearsal capabilities by October 2007; and
    — Creation of an initial Web-based delivery capability for operational mission planning and rehearsal by October 2005.
- Joint assessment and enabling capability milestones include:
    — Tracking of joint education and training experience of all DOD personnel by October 2005,
    — Linking of joint training to the Defense Readiness Reporting System network by March 2006, and
    — Ensuring that all DOD forces are trained prior to and during deployment by October 2007.

The DOD's Management Initiative Decision 906 indicates levels of funding during the FY 2003–2009 period for the training transformation effort: $86.4 million for the joint knowledge development and distribution capability; $1,121.7 million for JNTC; and $118.9 million for the joint assessment and enabling capability. This effort will provide a common architecture for linking test and training ranges.

The major training centers to be linked during the FY 2003–FY 2005 period are as follows:[33]

- U.S. Army National Training Center, Fort Irwin, California;
- Joint Readiness Training Center, Fort Polk, Louisiana;
- Fort Bliss Exercise Roving Sands training range, Fort Bliss, Texas;
- U.S. Navy Fleet East training area, Norfolk, Virginia;
- U.S. Navy Fleet West training area, San Diego, California;
- U.S. Air Force Nellis test and training ranges, Nellis Air Force Base, Nevada; and
- U.S. Marine Corps Twenty-Nine Palms range, Twenty-Nine Palms, California.

Future plans include linking training ranges worldwide and providing a deployable capability by October 2007.

## 3.6 GLOBAL INFORMATION GRID

OSD, and particularly the ASD(NII), has undertaken both policy and program initiatives to promote what they call "netcentricity." These activities are centered on the concept of a GIG. They will both facilitate and constrain DON's implementation of FORCEnet.

### 3.6.1 Policy Initiatives

Policy initiatives are intended to ensure interoperability and to promote a Services-oriented architecture. Table 3.1 lists some of the documentation related to these initiatives.

The Net-Centric Checklist, summarized in Table 3.2, will be used in OSD program reviews to ensure that the Services and combat support agencies support network-centricity. Several policy memoranda[34] make clear the OSD's commitment to this transformation. Programs will be classified according to their degree of conformity; nonconforming programs will be targeted for termination by prohibiting their further acquisition and deployment and by decrementing the funding for their maintenance.

---

[33]Deputy Secretary of Defense. 2003. Management Initiative Decision 906, U.S. Department of Defense, Washington, D.C., January, p. 13.

[34]These include DOD Directive 8100.1, "Global Information Grid Overarching Policy," September 19, 2002; Deputy Secretary of Defense Memorandum, "Global Information Grid Enterprise Services: Core Enterprise Services Implementation," November 10, 2003; and Assistant Secretary of Defense for Networks and Information Integration Guidance Memorandum, "Global Information Grid Enterprise Services," February 2004.

TABLE 3.1 Key Documentation Related to the Global Information Grid (GIG)

| Document | Purpose |
|---|---|
| Net-Centric Operations and Warfare Reference Model | Serves as a reference for network-centricity in the development of Department of Defense (DOD) architectures and in DOD oversight processes—describes enterprise-level activities, services, technologies, and concepts. |
| DOD Net-Centric Data Strategy | Defines a vision for data management within the DOD, emphasizing visibility and accessibility. |
| GIG Core Enterprise Services (CES) Strategy | Defines each of the CES, describes the technical capabilities that will be delivered by each of the CES, and presents a strategy for the evolution of capabilities. |
| Joint Technical Architecture, Version 6.0 | Delineates mandatory standards and guidelines, lists emerging, network-centric standards and guidelines as reference material for acquisition. |
| Net-Centric Checklist, Version 2.1 | Serves as a guide to understanding network-centric attributes required for programs to move into the GIG network-centric environment. |
| GIG Architecture, Version 2.0 | Describes the GIG architecture. |
| Transformational Communications Architecture (TCA) | Describes the TCA. |

### 3.6.2 Components of the Global Information Grid

Table 3.3 describes the eight investment areas leading to the realization of the GIG. The first seven are programs. The eighth, Horizontal Fusion, is a portfolio of experiments and demonstrations managed by the ASD(NII).

Each of the seven development programs has one or more executive agents drawn from among the Services and combat support agencies. One risk, not shown in the table, is that funding for Service-managed programs must compete within their own Service and at the JROC.

### 3.6.3 Implications for FORCEnet

When fully implemented, the GIG will be capable of performing considerable "heavy lifting" for FORCEnet. Provided that DON buys suitable terminals and sites them properly, TSAT promises to provide T-1 level (1.5 Mbps) service to 1-foot apertures on the move, and much higher capacities to larger antennas. Optical exfiltration and backhauling of airborne surveillance data will free unmanned air vehicles from a line-of-sight tether to ships and will make possible

TABLE 3.2 Summary of the Net-Centric Checklist Used by the Office of the Secretary of Defense in Program Reviews for the Services and Combat Support Agencies

| Title | Description |
|---|---|
| Internet Protocol (IP) | Data packets are routed across the network, not switched via dedicated circuits. |
| Secure communications | Communications are encrypted initially for core network; goal is edge-to-edge encryption. |
| Only handle information once (OHIO) | Data are posted by authoritative sources and are visible, available, usable, so as to accelerate decision making. |
| Post in parallel | Data are available on the network as soon as they are created. |
| Smart pull (versus smart push) | Applications encourage discovery; users can pull data directly from the network. |
| Data-centric | Data are separate from applications; applications talk to each other by posting data. |
| Application diversity | Users can pull multiple applications to access the same data or choose the same applications (e.g., for collaboration). |
| Dynamic allocation of access | Trusted accessibility exists for network resources (data, services, applications, people, collaborative environment, and so on). |
| Quality of service | Data have timeliness, accuracy, completeness, integrity, and ease of use. |

NOTE: GIG Arch v2, Global Information Grid (GIG) Architecture, Version 2.0.

smaller ship crews by exploiting the data at combatant command headquarters and in the continental United States.

The GIG-BE will simplify the work of the Naval Computer and Telecommunications Area Master Station (NCTAMS) and the Network Operations Center System and reduce the load on satellite communications. JTRS will support force composability both through making the wideband networking waveform available to all forces and by interoperating with legacy radios. NCES promises to reduce nonrecurring cost in acquiring new software capabilities and to simplify the maintenance and operation of deployed systems. Joint command and control promises interoperability among Service command-and-control systems—a promise that the Global Command and Control System (GCCS) never quite kept.

| Metric | Source |
| --- | --- |
| Net-Centric Operations and Warfare Reference Model (NCOW RM) compliance. | NCOW RM, GIG Arch v2, IPv6 Memos (June 9, 2003, and Sept. 29, 2003) |
| Transformational Communications Architecture (TCA) compliance. | TCA |
| Reuse of existing data repositories. | Community-of-interest policy (to be determined). |
| NCOW RM compliance. Data are tagged and posted before processing. | NCOW RM, DOD Net-Centric Data Strategy (May 9, 2003) |
| NCOW RM compliance. Data are stored in public space and advertised (tagged) for discovery. | NCOW RM, DOD Net-Centric Data Strategy (May 9, 2003) |
| NCOW RM compliance. Metadata are registered in DOD Metadata Registry. | NCOW RM, DOD Net-Centric Data Strategy (May 9, 2003) |
| NCOW RM compliance. Applications posted to the network and tagged for discovery. | NCOW RM |
| Access is assured for authorized users, denied for unauthorized users. | Security information assurance policy (to be determined). |
| Being network-ready is the key performance parameter. | Service-level agreements (to be determined). |

The Distributed Common Ground/Surface Systems (DCGS) will make data from the sensors of all Services and agencies available in fused and actionable form.

The network-centric policies of the OSD provide external leverage for making new naval platforms network-centric. Nevertheless, the GIG presents challenges as well as opportunities for FORCEnet. Much of the GIG philosophy is based on a promise that communications bandwidth will no longer be a constraint. That assumption makes a transition to an enterprise architecture within thin clients (those with limited processing and storage capability) attractive. Experience in the commercial sector indicates that a transition to an enterprise architecture may increase communications traffic by several orders of magnitude. On the other hand, the committee heard that many major ships have very limited

TABLE 3.3  Components of the Global Information Grid

| Component | Description | Availability[a] | Issues/Risk |
|---|---|---|---|
| Communications | | | |
| Transformational Satellite | Optical crosslinks and unmanned air vehicle uplinks; high-performance extremely high frequency. | After FY 2012 | Major space program with new technology. |
| GIG-BE | Global fiber network. | FY 2004 | Terrestrial only. |
| Joint Tactical Radio System | Software-based radios for legacy waveforms and new networks. | FY 2005–2007 for ground and air systems; FY 2009 for maritime systems. | Delivery time of maritime capability. |
| Crypto Xformation | Packet encryption for black core. | After FY 2005 | Overhead. Converting ships to black routing. |
| Services | | | |
| Network-Centric Enterprise Services | Common services for entire Department of the Navy enterprise. | FY 2008 | Performance and scalability. Dependence on perfect communications. |
| Command-and-Control Systems | | | |
| Distributed Common Ground/Surface Systems | Processing, sharing, and fusing of airborne collections. | After FY 2009 | Naval Fires Network incompatibility. |
| Joint Command and Control | Command and control for all Services; Global Command and Control System (GCCS) successor. | FY 2010 | Schedule. Needed GCCS-Maritime upgrades. |
| Integration | | | |
| Horizontal Fusion | Demonstrations of information sharing and fusion. | Variable | Incorporating successes into Programs of Record. |

[a]Expected fielding time frame.

bandwidth and are subject to frequent communications outages caused by antenna blockages. Until the performance of NCES is known, and until it is clear to what degree the promise can be kept, one cannot be sure that NCES will support FORCEnet.

NCES efforts, in the main, have to do with network infrastructure. Issues of information flow, content, management, and representation do not currently have a strong GIG focus. These issues are represented in various forms and for different purposes by communities of interest (COIs), which are conceived (at least initially) as traditional groupings (e.g., processes) from non-network experiences. Developments within these COIs are highly driven by the culture of preexisting associations and are therefore likely to revert to traditional standards rather than to converge on common network paradigms (e.g., common ontologies) unless properly focused. No compliance criteria for COI participation have been developed. Issues of information integrity drive the concern in this area. Without a cohering function, the COIs will develop differing conventions (from data dictionaries to processing architectures), and the ability to maintain consistency when enterprise sharing is required will diminish.

Issues raised in Chapter 1 relating to information integrity across the network, information management, and system performance measures are not visibly addressed in the GIG. Some joint efforts may be addressing these issues, but the committee did not encounter them.

Combat systems require an extremely high degree of integrity, and FORCEnet includes combat systems. Chapter 5 discusses the issue of separating combat loops from the rest of FORCEnet so as to avoid the expense of combat-certifying the entire FORCEnet as spirals are developed.

Transition to IPv6 and to black core with IP encryptors cannot be accomplished overnight. DON will have to devise a plan with staged implementation and procedures to allow interoperation between converted and unconverted units.

## 3.7 COALITION OPERATIONS

All major U.S. operations and the majority of exercises in the Joint Exercise Program involve coalitions. These exercises provide the major venues for experiments, tests, and demonstrations for developing coalition capabilities and interoperability. U.S. naval forces also participate in demonstrations, exercises, and operations scheduled and run by allies and coalition partners. The North Atlantic Treaty Organization (NATO) has formal approaches for interoperability development, whereas the United States has vigorous but less-formal processes for developing interoperability with its other allies and security partners. The development of CENTRIX (Combined Enterprise Regional Information Exchange) and associated standard operating procedures for coalition planning and operations in the U.S. Central and Pacific Commands has been the foundation for recent coalition operations. At the level of operational planning and coalition (blue) force tracking, FORCEnet will extend into coalition operations. Some allies, such as Japan and Korea, with Aegis capabilities, are working toward interoperability in missile and air defense at the level of fire control.

The NATO Summit held in Prague in 2002 resulted in a major reorganization: there are now two major NATO commands—Allied Command Operations (ACO), headquartered in Europe, and Allied Command Transformation (ACT), headquartered in North America. NATO is fully embracing this transformation process, under the leadership of ACT. A critical element in that transformation is the generation of the NATO Network-Enabled Capability (NNEC). The NATO secretary general has stated that "Allied Command Transformation will shape the future of combined and joint operations and will identify new concepts and bring them to maturity, and then turn these transformational concepts into reality—a reality shared by the whole NATO alliance."[35] ACT will incorporate into the NATO inventory those concepts that address the needs of the future allied operating environment. NNEC will strive to integrate systems from across the alliance, resulting in an interoperable system of systems.

To achieve an operational system of systems, NATO requires a methodology for developing that architecture. The methodology is to enable the creation of multifunctional, multilayered communications and information systems that are consistent with the overarching NNEC concept. Furthermore, the methodology intends to ensure interoperability by formalizing requirements and specifying intersystem standards. Within NATO, the Research and Technology Organization (RTO) identifies, conducts, and promotes cooperative research and information exchange that meets the military needs of the alliance. In addition, RTO has the ability to draw upon resources across the alliance to carry out that task.

U.S. efforts to develop network-centric capabilities are leading allied and coalition efforts. Through routine interaction with allies and coalition partners, U.S. naval forces are well positioned to further FORCEnet implementation in this context.

## 3.8 CHALLENGES IN BRIDGING NETWORK-CENTRIC CONCEPTS AND JOINT CAPABILITIES

One of the most effective force transformations in recent history resulted from the Navy's creating Submarine Development Group TWO in 1949, with the mission "to solve the problem of using submarines to detect and destroy enemy

---

[35]Remarks by the Secretary General of NATO, Lord George Robertson, at the ceremony establishing the new NATO Transformation Command in Norfolk, Virginia, June 19, 2003. Available at www.defenselink.mil/news.Jun2003/N06192003_200306193.html. Accessed July 24, 2004.

submarines."[36] In 1949, the U.S. submarine force had no capability to sink a submerged submarine. By 1969, as a result of Submarine Development Group TWO exercises, analyses, and continuous tactics and technology development, the U.S. submarine force had become the dominant antisubmarine capability in the world. This success derived from "a willingness to innovate, close and open ties to the technical community, unblinking candor in performance analysis, dedicated organic submarines focused on development, top-notch personnel, military and civilian, and a strong, clear mission focus."[37]

The defense planning, joint capabilities integration and development, joint concept development and experimentation, JBMC2, joint testing, joint training, GIG development and acquisition processes, and coalition considerations described in this chapter are all separate activities with little interaction. Proposals to organize these activities around mission areas have proven difficult to implement for the large bureaucracies involved in each activity. Though efforts such as JBMC2 are striving to develop architectures around joint mission threads,[38] the relationships between these architectures and network-centric enterprise architectures have yet to be illustrated. Absent an integrated program of concept development, experimentation, technology insertion, and system testing in joint and coalition exercises, it is highly uncertain that the challenges involved in transforming to network-centric enterprises and FORCEnet can be resolved, particularly at the tactical level where quality-of-service and latency issues are acute.

## 3.9 FINDINGS AND RECOMMENDATIONS

### 3.9.1 Findings

Following are the findings of the committee with respect to joint capability development and DOD network-centric plans and initiatives:

• The leadership of the Navy and the Marine Corps is committed to the development of capabilities that will enable them to operate as components of a joint force with network-centric attributes. All recent combat operations have

---

[36]"Submarine Warfare and Tactical Development: A Look—Past, Present and Future." 1999. In *Proceedings of the Submarine Development Group TWO and Submarine Development Squadron TWELVE 50th Anniversary Symposium, 1949–1999*, held at U.S. Naval Submarine Base, Groton, Connecticut, May, p. 124.

[37]"Submarine Warfare and Tactical Development: A Look—Past, Present and Future." 1999. In *Proceedings of the Submarine Development Group TWO and Submarine Development Squadron TWELVE 50th Anniversary Symposium, 1949–1999*, held at U.S. Naval Submarine Base, Groton, Connecticut, May, pp. 21–22.

[38]A description of mission threads is given in Chapter 4, Section 4.7.1.

been joint, and the extent of joint interaction in military operations is only likely to increase. Thus, FORCEnet operational and materiel capabilities must be developed in a joint context.

• The DOD requirements prioritization and acquisition, concept development and experimentation, testing, and training are in flux. Further refinements in these processes should be anticipated. It is critical that FORCEnet implementation couple into these larger DOD-wide processes. Aligning FORCEnet implementation with the guidance from the OSD regarding network-centric operations and warfare, increased support to combatant commanders, JBMC2, and the GIG will be challenging, but also could represent opportunities.

— The new Joint Programming Guidance from the OSD will direct a greater portion of DON resources toward OSD and combatant commander priorities. If the new process works as intended and FORCEnet is perceived as providing joint capabilities responsive to combatant commander needs, FORCEnet is less likely to have to compete for funding within the available discretionary funds of the Navy and Marine Corps that will have been reduced as a result of mandated spending on joint capabilities. Whether the effect on FORCEnet is positive or not remains to be seen.

— There is no set of future concepts for joint operations with adequate detail to inform and guide the Navy and Marine Corps in developing their concepts for participating in joint operations. Joint efforts to date based on the JCIDS have been largely concerned with very broad conceptual development. The Joint Staff has recently initiated work on a set of Joint Integrating Concepts that may provide the required specificity.

— The JBMC2 roadmap treats FORCEnet as providing the single integrated maritime picture. The roadmap is meant to drive program priorities and schedules, presenting a potential challenge to FORCEnet spiral development. FORCEnet's inclusion in the JBMC2 roadmap will create pressures for FORCEnet to synchronize development with other Service and joint efforts and programs, such as the Army's Future Combat System (FCS), the Air Force Command and Control Constellation, and the joint DCGS and the related Joint Fires Network (JFN)/Tactical Electronic Surveillance developments.

— Joint lessons learned from OIF and the major combat operations phase of OIF are affecting defense planning priorities. They emphasize joint action facilitated by shared awareness and interoperability. Ongoing operations in Iraq are pressing the OSD and the Services to adopt new approaches for rapidly fielding capabilities to address the challenges that deployed forces are facing, many of which involve networking and shared awareness. Since the lessons principally involve joint interoperability issues that affect ground operations, this effort may direct a significant portion of FORCEnet development toward providing remedies needed for Navy and Marine Corps ground operations.

— The JFCOM joint experimentation process is beginning to transition from being based on JFCOM-originated concepts focused at the operational level of war to becoming a broader process more widely serving the needs of the joint community and Services. Further progress in this direction is necessary, with particular attention on the tactical level of warfare, given the growing joint inter-action at that level evident in recent conflicts. In participating in JFCOM experi-mentation activities, the Navy and Marine Corps need to keep their activities focused so that they do not become overwhelmed by the much greater JFCOM experimentation resources.

— While JFCOM is the executive agent for joint experimentation, the regional combatant commands are becoming an increasing focus for joint con-cept development and experimentation. The fleet commands are a natural vehicle for interacting with the combatant commands in this regard, as has been the case such as in the interaction of the Pacific Fleet and its components with PACOM.

— The OSD is trying to create the functionality of the Navy DEP in a JDEP. As JDEP develops, it has the potential to provide joint interoperability and integration infrastructure for FORCEnet in a manner similar to the way that the Navy DEP provides it for the fleet. The spiral development of FORCEnet will require capabilities similar to those developed for JDEP. The Navy is positioned to influence this development in ways that support Sea Trial and FORCEnet implementation. The Navy DEP and JDEP, currently focused on system interoper-ability, will need to evolve to support the network-centric aspects of the GIG.

— The OSD-led training transformation involves significant investment that could be employed for FORCEnet development. FORCEnet implementation can leverage the DOD investment in training infrastructure in several ways, including distributed training and education of naval personnel as FORCEnet capabilities develop and through the development of FORCEnet capabilities in joint training exercises. The training transformation is meant to provide a com-mon architecture for live, virtual, and constructive training and embedded train-ing in major acquisitions programs that allows systems to link immediately into the global joint training infrastructure. Training with forces from other Services is expected to become routine as the training ranges become linked and deploy-ment schedules are aligned. The joint assessment capability also provides a means for documenting capability enhancements provided by FORCEnet. Documented capability enhancements can help justify adaptive expenditures that support short spiral times.

• Efforts by the ASD(NII) have created architectures, reference models, standards, and new paradigms (e.g., task, post, process, use (TPPU); and only handle information once (OHIO)) to drive enterprise network operations. The sufficiency or completeness of these directions is unknown.

— Enterprise services have been described as a paradigm for network operations. No services-oriented architecture of this scale has been attempted.

— The GIG is driven principally by technological considerations and network-centric theories and may not satisfy all warfighting needs.

•   FORCEnet implementation is finding the same challenges as those that the DOD faces in strengthening jointness and migrating to network-centric concepts and systems. This process involves many open questions that require careful design, experimentation, and growing experience to resolve. A successful FORCEnet implementation strategy has the potential to be a model for realizing network-centric capabilities across DOD.

### 3.9.2 Recommendations

Based on the findings presented above and on the issues described in this chapter, the committee recommends the following:

•   **Recommendation** for NETWARCOM, NWDC, and MCCDC: Continue to work with JFCOM to broaden its experimental perspective, with particular emphasis on joint operations at the tactical level. If necessary for these organizations to maintain focused commitment in the face of far larger JFCOM resources, the CFFC, and Commanding General, MCCDC, should provide guidance on the issues to be addressed and the partitioning of naval involvement in JFCOM, regional combatant commands, and Service concept development and experimentation activities.

•   **Recommendation** for the fleet commands and Marine Expeditionary Forces (MEFs): Build on current interactions with regional combatant commands in order to grow the relationship between naval and joint concept development and experimentation. This means ensuring both that naval concepts are properly embodied in joint concepts and that they reflect the needs of the joint concepts. Combatant command exercises should be used as a principal vehicle for exploring and refining the concepts. This responsibility could require that the fleets devote more resources to concept development.

•   **Recommendation** for NETWARCOM and MCCDC with technical support from such organizations as the Space and Naval Warfare Systems Command (SPAWAR) and the Office of Naval Research (ONR): Undertake a series of naval mission-based analyses to understand the technical limits to achieving network-centric operational concepts and identify approaches for dealing with potential operational degradations in network capabilities. Such analyses should indicate where reliance on more "traditional" capabilities (e.g., the use of localized versus distributed services) may still be necessary, and where increased attention to network path diversity and node heterogeneity is needed to reduce network vulnerability. These results should be shared with other Services and the joint community to increase the understanding of the limits on joint operations.[39]

---

[39]The mission thread analysis planned in conjunction with the FORCEnet baseline assesment could represent a start of the necessary analyses.

- **Recommendation** for the Deputy Chief of Naval Operations (DCNO) for Warfare Requirements and Programs (N6/N7) and the Deputy Commandant of the Marine Corps for Plans, Policies, and Operations (DCMC(PP&O)): Work to articulate clearly how FORCEnet capabilities pertain to joint operations and satisfy the needs of combatant commanders. In the context of the Joint Defense Capabilities Process and JCIDS, this line of argument will strengthen programs providing FORCEnet capabilities in the budget process. While assertions of the joint nature of FORCEnet capabilities have frequently been made by the Navy and Marine Corps in general terms, the committee has not seen any detailed analyses working through the arguments.

- **Recommendation** for the fleet commands and MEFs: Work with the combatant commands to which they are assigned in order to understand and feed into the naval requirements process the capabilities needed by the combatant commands from naval forces. The CFFC, and Commanding General, MCCDC, would act as the intermediaries for feeding this information from the fleets and MEFs into the program planning processes of the Navy and Marine Corps.

- **Recommendation** for the N6/N7 and the Marine Corps Director for Command, Control, Communication, Computers, and Intelligence (C4I): Adopt a prudent course with respect to joint GIG programs, endorsing the further development of these programs but also requiring a clear and continuing assessment of their technical and programmatic progress. In this context, the N6/N7 and the Director, C4I, should clearly understand the limits of applicability of network-centric capabilities, especially at the tactical level.

- **Recommendation** for the N6/N7 and N8, and the Deputy Commandant of the Marine Corps for Programs and Resources (DCMC(P&R)): Articulate programmatic strategies, updated on an annual basis, for leveraging progress and accommodating developments in joint GIG programs. This strategy should lay out approaches for developing the necessary complementary naval capabilities (e.g., terminals, antennas) and describe technical and programmatic alternatives corresponding to the status of joint programs—that is, whether they have remained on schedule, slipped, or failed to meet their objectives. The strategy should also indicate how to leverage joint GIG capabilities as they become available. While some such capabilities will not be deployable for many years (e.g., the TSAT), others will be available in the near term (e.g., initial releases of NCES, Horizontal Fusion services).

- **Recommendation** for the Assistant Secretary of the Navy for Research, Development, and Acquisition (ASN(RDA)) with support from the program executive officer (PEO) C4I & Space, the PEO Space Systems, SPAWAR, and the Marine Corps Systems Command (MARCORSYSCOM): Track and provide input to the technical development of joint GIG programs to ensure that as these programs evolve, they continue to satisfy naval needs. This objective is best accomplished through naval participation in the programs. The ASN(RDA) should build on current naval participation to ensure that the involvement remains substantive and is across all major GIG programs. The ASN(RDA) should also see that the proper operational perspective (e.g., through the involvement of NETWARCOM) is brought to bear in this activity.

• **Recommendation** for the N6/N7, the ASN(RDA), and the MARCOR-SYSCOM: Fully impose the network-centric criteria mandated by the ASD(NII), in the development and execution of naval programs, subject to any necessary refinement of these criteria. Since the criteria are in their initial use now, the N6/N7, the ASN(RDA), and MARCORSYSCOM should work with the ASD(NII) to refine these criteria as necessary, prior to their full imposition. The use of these criteria will further strengthen related internal policies of the Navy and Marine Corps. Furthermore, if the ASD(NII) network-centric reviews gain strong influence in the DOD budget process, meeting the criteria will be necessary to ensure adequate funding of programs.

• **Recommendation** for the N6/N7 and the ASN(RDA):[40] Work with the OSD and the other Services to develop a better understanding of, and eventually to develop guidelines and principles for, how the numerous architectures being developed in DOD can be effectively integrated. Particular attention is necessary at the tactical level of warfare, since architectural development for the GIG has not explored that level to a significant extent. The N6/N7 and the ASN(RDA) would be supported by SPAWAR, NETWARCOM, and the MARCORSYSCOM in this work. Interaction with the combatant commands (particularly the JFCOM and STRATCOM) and combat support agencies (particularly the Defense Information Systems Agency (DISA)) would also be required.

• **Recommendation** for the ASN(RDA) and the CFFC: Coordinate with the Director of Operational Test and Evaluation and the DISA to leverage the DOD investment in the JDEP while expanding the Navy DEP to accommodate the spiral development of FORCEnet capabilities.

• **Recommendation** for the ASN(RDA) and the DCNO for Fleet Readiness and Logistics (N4): Coordinate with the Under Secretary of Defense for Personnel and Readiness to exploit DOD investments in Training Transformation to support FORCEnet development. The committee recommends that CFFC and Sea Trial operational agents schedule fleet battle experiments in exercises employing the joint national training capability.

• **Recommendation** for CFFC and NETWARCOM: Interface with NATO ACT and the NATO RTO to foster interoperable coevolution of the capabilities of FORCEnet and NATO Network Enabled Systems. Commanders of the Pacific Fleet and Fifth Fleet, as naval component commanders respectively in PACOM and U.S. Central Command, should coordinate the specific requirements for the coevolution of FORCEnet capabilities in their theaters.

---

[40]If a director of FORCEnet were appointed, that individual would be the appropriate party to lead naval efforts in effecting this and the previous recommendations.

# 4

# Coevolution of FORCEnet Operational Concepts and Materiel

Simply grafting new technology to old processes will not work. To fully leverage the advantages technology brings, we must speed our process of innovation
and co-evolve concepts, technologies, and doctrine.[1]

## 4.1 INTRODUCTION

Chapter 2 introduced the notion of integrating spiral developments of operational constructs and materiel architecture and technology. The operational-con-
struct spiral involves the coevolution of doctrine, organization, training, leadership and education, personnel and facilities (nonmateriel solutions), with changes
in materiel to take advantage of emerging technology and dynamic challenges.

However, the traditional process for the acquisition of large, capital-inten-
sive systems—ships, submarines, aircraft, and spacecraft—so-called ACAT I
programs,[2] is a linear process dominated by large up-front investments in time
and resources to ensure that the relatively small number of systems procured are
the best and most cost-effective available at the time. That process can be summarized as follows:[3]

---

[1]ADM Robert J. Natter, USN. 2003. "Sea Power 21 Series, Part VIII: Sea Trial: Enabler for a
Transformed Fleet," *U.S. Naval Institute Proceedings*, November, p. 62.

[2]The ACAT designations (I, II, III, and so on) are established for all the military Services by DOD
Instructions 5000.1 and 5000.2 and their Service-specific supplements.

[3]A useful review of the acquisition process for ships can be found in Robert S. Leonard, Jeffery A.
Dreener, and Geoffrey Summers, 1999, *The Arsenal Ship Acquisition Process Experience*, RAND,
Santa Monica, California.

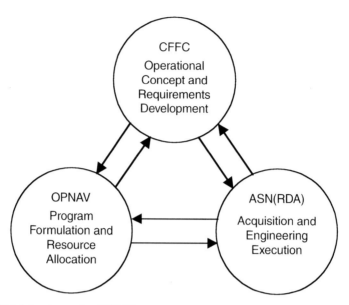

FIGURE 4.1  Implementing FORCEnet.

1. Conduct studies and experimentation to refine capability needs and potential solutions;
2. Generate functional capability requirements and priorities;
3. Gain approval of capabilities through the DOD budget;
4. Establish the program, its governance, and its milestones;
5. Conduct design and feasibility studies and establish the initial design;
6. Award contracts for initial, low-rate production; and
7. Reanalyze the design and contracting and award follow-on contracts for full production.

The challenge in implementing FORCEnet is to make this linear process highly iterative and integrate it with concept development, as suggested by Figure 4.1.

The authority for each of the three major FORCEnet implementation activities is indicated in the diagram: the CFFC for operational concept and requirements development; the ASN(RDA) for acquisition and engineering execution; and the OPNAV for program formulation and resource allocation. However, the responsibilities for these activities are even more distributed, as indicated in Tables 4.1 ("Navy FORCEnet Implementation Responsibilities") and 4.2 ("Marine Corps FORCEnet Implementation Responsibilities"). Although some of this diffusion is required by law, successful implementation of FORCEnet will re-

TABLE 4.1 Navy FORCEnet Implementation Responsibilities

| Functional Area | Organization | Responsibilities |
|---|---|---|
| Operational concepts | CFFC | Oversee concept development and experimentation (CD&E). |
| | Second Fleet | Conduct CD&E for Sea Strike and Sea Basing. |
| | Third Fleet | Conduct CD&E for Sea Shield. |
| | NETWARCOM | Conduct CD&E for FORCEnet and ensure alignment with joint concepts. |
| | NWDC | Coordinate CD&E. |
| Requirements | CFFC | Lead development of fleet operational requirements. |
| | Second Fleet, Third Fleet | Determine requirements for Sea Power 21 pillars and relate them to needed FORCEnet capabilities. |
| | NETWARCOM | Determine FORCEnet requirements. |
| Programs and resources | OPNAV N6/N7 | Validate and prioritize FORCEnet requirements for program development and coordinate with other warfare area sponsors. |
| | OPNAV N8 | Assess programs for resourcing and requirements. |
| Acquisition | ASN(RDA) | Oversee acquisition of all FORCEnet capabilities and ensure compliance with FORCEnet architecture. |
| | PEOs | Oversee program execution in area of jurisdiction. |
| | CNR | Oversee Navy science and technology development for FORCEnet capabilities. |
| Engineering | SPAWAR | Develop FORCEnet architecture and function as FORCEnet chief engineer. |
| | NAVSEA, NAVAIR | Develop architectures for Sea Power 21 pillars. |
| | PEOs | Apply architectures in program execution. |

NOTE: Acronyms are defined in Appendix C.

quire close coordination of and collaboration among these activities. This chapter examines the activities and the prospects for improving their coordination.

## 4.2 COEVOLUTION OF OPERATING CONCEPTS AND TECHNOLOGY INTO WARFIGHTING CAPABILITIES

The enhanced capabilities of Sea Power 21 are made possible by the technical capabilities of the FnII and of the systems that it interconnects. More importantly, however, FORCEnet implementation requires the development of new operational processes—concepts of operations and tactics, techniques, and procedures (TTPs)—that take advantage of the new FnII capability if advances are to be achieved in the naval warfighting capabilities represented in Sea Strike, Sea Shield, Sea Basing, and EMW. Coevolution of technical-capabilities development and operational-concept development is the process by which change in one can be synchronized with change in the other.

TABLE 4.2  Marine Corps FORCEnet Implementation Responsibilities

| Functional Area | Organization | Responsibilities |
|---|---|---|
| Operational concepts | MCCDC, MCWL | Oversee concept development and experimentation. |
| Requirements | MCCDC | Lead long-term USMC requirements. Lead command element requirements. |
| | DCMC(PP&O) | Lead ground combat element requirements. |
| | DCMC(Aviation) | Lead aviation requirements. |
| | DCMC(I&L) | Lead logistics and facilities requirements. |
| Programs and resources | DCMC(P&R) | Serve as program and resource sponsor for all USMC programs. |
| Acquisition | ASN(RDA) | Oversee acquisition of all FORCEnet capabilities. |
| | MARCORSYSCOM | Oversee and execute all USMC programs. |
| Engineering | MARCORSYSCOM | Conduct engineering development for USMC programs. |

NOTE: Acronyms are defined in Appendix C.

With such coevolution, when improved capabilities are deployed, the operational concepts initially employed are those currently in existence that are most closely related to the new capability. The operational concepts are adjusted or changed only after experience with the new equipment is gained in the operational environment. For example, when the F/A-18 Hornet was first introduced to fleet operations in the post–Vietnam War era, flight profiles used in large-airwing strike packages (known as Alpha strikes) required all strike aircraft, regardless of type, to rendezvous and fly together in a large formation to the target for mutual protection en route and to facilitate a coordinated, near-simultaneous attack. This worked reasonably well when the aircraft mix included A-6s, A-7s, and F-14s, because none suffered a significant performance (fuel) penalty for the air speeds and altitudes used en route to the target. Once the F/A-18 took the place of the A-7 in the strike formation, the operational concept for Alpha strikes had to change to allow the F/A-18s, for fuel efficiency, to fly a higher, faster flight profile than that of the A-6s and F-14s. Designating a fixed time on target provided the means for coordinating the attacks, while the F/A-18s flew a different speed and altitude profile to maximize aircraft performance.

Just as introducing new capability stimulates change in operational concepts, changes in the operational environment can drive change as well. An example of such an interaction is found in the contrast between Operation Desert Storm and more recent operations in Afghanistan and Iraq. In Operation Desert Storm, air operations were scheduled via a serial process, the ATO, based on a 72-hour planning and execution cycle. Targets were picked early in the cycle and attacked in the last 24 hours of the cycle. Three 72-hour cycles ran concurrently, each staggered from the next by 24 hours; thus, as target development was beginning for one, another was in the middle of ATO production, and the third was in execution on the same day. Except for close-air-support missions, specific targets were assigned ahead of time with the publication of the ATO, and they generally remained unchanged as aircrew and sortie assignments were made, flight plans were briefed, aircraft were manned and launched, and weapons were delivered in the execution phase of each 72-hour ATO cycle.

The formality of the 72-hour ATO cycle works well against fixed targets in a war of attrition such as that accomplished by air operations in Desert Storm prior to the initiation of ground action. But the need for more flexibility became apparent with the initiation of Operation Enduring Freedom in Afghanistan; in this conflict the operational environment initially precluded basing aircraft within reasonable distances of the targets, and the targets were more fugitive. Mission flight times during Operation Enduring Freedom increased by three to four times over those of Desert Storm, and targets were more transient. In many cases, special operations forces on the ground identified time-sensitive targets that required dynamic pairing of weapons to targets in real time—otherwise the opportunity was lost. To meet the need, the nature of the ATO changed to make more extensive use of a concept called flex targeting, in which some aircraft are launched without target assignments.

This type of change in operating concept, brought about by a change in the operating environment, can have significant impact on the FnII capabilities that are needed. In this example, the location for the delivery to the aircrew of up-to-date battlefield intelligence and targeting information, including recent imagery, moves from the preflight briefing room on deck to the aircraft cockpit airborne somewhere en route to the target. The need for a means to receive—at over-the-horizon distances and display in usable form—the needed intelligence and targeting information places a new requirement on FnII.

The foregoing examples involved new concepts for the use of already-deployed materiel. The coevolution of concepts and materiel may involve experimenting with prototypes so that the value of the combination of new materiel capabilities and concepts for their use can be evaluated as each is refined. The Navy has formalized this process under the name Sea Trial. Figure 4.2 is drawn from the instruction that describes the Sea Trial process. This process begins with the generation of concepts in response to warfare challenges, as discussed in the next section.

## 4.3  CONCEPT DEVELOPMENT

### 4.3.1  Concept Hierarchies

The Navy's approach to concept development, applied by the Concepts Development Department of NWDC in partnership with the fleet and the Marine

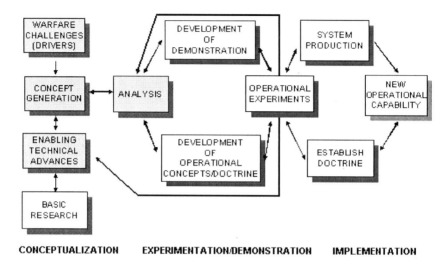

FIGURE 4.2 Notional Sea Trial capabilities development process. SOURCE: ADM Robert J. Natter, USN. 2003. "Sea Power 21 Series, Part VIII: Sea Trial: Innovation Enabler for a Transformed Fleet," *U.S. Naval Institute Proceedings*, November, p. 62.

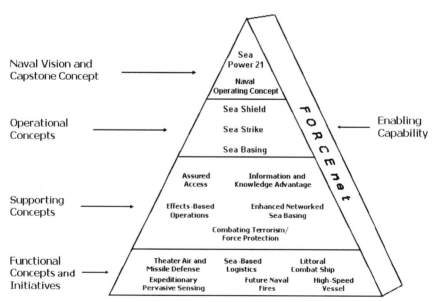

FIGURE 4.3 Naval concept hierarchy. SOURCE: Navy Warfare Development Command, Newport, R.I.

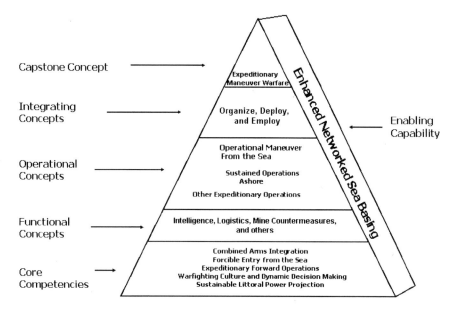

FIGURE 4.4 Marine Corps concept hierarchy. SOURCE: Marine Corps Combat Development Command, Quantico, Va.

Corps, is shown in Figure 4.3. FORCEnet is shown as the enabling capability that spans every level of the concept hierarchy. The Marine Corps has a similar concept hierarchy, with the capstone concept EMW, as shown in Figure 4.4.

The Naval Operating Concept flows from the vision of Sea Power 21 and the strategy of Sea Power 21 and Marine Corps Strategy 21, from which the supporting concepts are developed. Naval forces support the pillars of Sea Power 21 (Sea Shield, Sea Strike, and Sea Basing) through specific concepts as well as through traditional naval capabilities.

Figures 4.3 and 4.4 reflect the FORCEnet-enabled capabilities for naval concepts. The Navy specifically intends to support the Sea Power 21 pillars through Mission Capability Packages (MCPs), listed in boxes in the subsections below, whereas the Marine Corps provides specific support to the MCPs in the form of embarked forces, fires, equipment, and capabilities. In addition, the Marine Corps provides general and specific support to Sea Power 21 pillars in the form of task-organized Marine Air-Ground Task Forces (MAGTFs).

### 4.3.2 Navy Concept Development

NWDC coordinates concept development for the Navy, but the CFFC has assigned concept-development responsibility for the Sea Power 21 pillars to operational agents—the Second Fleet for Sea Strike and Sea Basing and the Third

Fleet for Sea Shield—and has assigned concept-development responsibility for FORCEnet to the NETWARCOM. Note that distinguishing FORCEnet from the Sea Power 21 pillars risks confining "FORCEnet concept development" to the FnII.

Nevertheless, NETWARCOM is engaged in a collaborative effort with many other participants—the Navy War College, the NWDC, the Net-Centric Warfare Directorate of the DCNO for Warfare Requirements and Programs (N71, formerly N61),[4] MCCDC, SPAWAR, numbered fleet commanders, and Warfare Centers of Excellence—to provide a functional-level concept that describes how future joint and combined network-centric capabilities may be used to facilitate and enhance naval operations in the 2015–2020 time frame. The functional concept for FORCEnet is intended to support the development process for FORCEnet transformational requirements, the development of the FORCEnet operational architecture, and CD&E. The FORCEnet concept is to evolve to serve as a coherent unifying concept that enables Sea Strike, Sea Shield and Sea Basing. As of this writing, a final draft for senior leadership review was planned for June 30, 2004.[5]

### 4.3.3  Marine Corps Concept Development

Concept development in the Marine Corps is a continuing process. It occurs as the nature of warfare changes, trends are identified, capabilities are assessed, and concepts are written and requirements validated. The commanding general of MCCDC at Quantico, Virginia, is formally tasked with concept development for the Marine Corps. Although ideas and initiatives for concepts may originate from numerous sources, the Expeditionary Force Development Center at Quantico actually writes and publishes Marine Corps concepts.

As shown in Figure 4.4, the Marine Corps has several types of concepts. EMW is the Marine Corps capstone concept. EMW is the union of core competencies, maneuver warfare philosophy, expeditionary heritage, and the concepts by which the Marine Corps organizes, deploys, and employs forces. Integrating concepts for MAGTF organizations are broad-based in nature and define capabilities, organizational structures, and force maneuver options. Operational concepts for Ship-to-Objective Maneuver are based on the Marine Corps warfighting philosophy of maneuver warfare: that is, seeking to shatter enemy cohesion through a series of rapid, violent, and unexpected actions that create turbulent and deteriorating situations with which the enemy cannot cope. Functional concepts

---

[4]The Net-Centric Warfare Directorate is the program sponsor for space, naval, and shore communications, networks, command and control, intelligence surveillance and reconnaissance, and intelligence oversight.

[5]Naval Network Warfare Command. 2004. White paper: "FORCEnet Concept Development," Norfolk, Va., February 24.

---

**BOX 4.1**
**The Four Naval Capability Pillars of Sea Power 21 and
Their Mission Capabilities**

| Sea Shield | Sea Strike | Sea Basing | FORCEnet |
|---|---|---|---|
| • Force Protection | • Strike | • Deployment and | • Intelligence, |
| • Surface Warfare | • Fire Support | Employment of | Surveillance, |
| • Undersea Warfare | • Maneuver | Naval Forces | Reconnaissance |
| • Theater Air and | • Strategic | • Provision of | • Common |
| Missile Defense | Deterrence | Integrated Joint | Operational and |
| | | Logistics | Tactical Pictures |
| | | • Pre-positioning of | • Communications |
| | | Joint Assets Afloat | and Data Networks |

---

address capabilities in specific areas (such as logistics) and tend to be more technical than operational concepts are.

Before being formally adopted, concepts are exposed to rigorous examination. As concepts are being developed, they are subject to war gaming, experimentation, modeling and simulation, tabletop seminars with subject-matter experts, Marine Corps Unit review, and operational evaluation by fleet forces. Core competencies are signature characteristics of Marines and the Marine Corps.

### 4.3.4 The Navy Pillar Concepts and Capabilities

The Navy is in the process of further refining and defining operating concepts for the three Sea Power 21 pillars of Sea Shield, Sea Strike, and Sea Basing. The three pillars, together with FORCEnet, constitute the four Naval Capability Pillars (NCPs) (see Box 4.1). Each NCP is further divided into MCPs[6] that relate to the broad missions that the NCP is to address (see the subsections below). Each MCP contains several specific capabilities that must be realized to some level in each deploying Joint Maritime Force Package.[7]

Note in Box 4.1 that the FORCEnet NCP comprises little more than the FnII. Even the Intelligence, Surveillance, and Reconnaissance (ISR) MCP does not include all ISR capabilities, many of which are organic to platforms in other NCPs. One consequence of this narrow definition of the FORCEnet NCP is to

---

[6]The term "Mission Capability Package" has a different meaning in the Office of the Chief of Naval Operations, as discussed in Section 4.5.1.

[7]The composition of deployable force packages is discussed in VADM Michael Mullen, USN, 2003, "Sea Power 21 Series, Part VI: Global Concept of Operations," *U.S. Naval Institute Proceedings*, April.

narrow the role of the Naval Network Warfare Command, which is the FORCEnet operational agent, in devising concepts for network-centric operations.

### 4.3.4.1 Sea Shield

The Sea Shield NCP has four Mission Capability Packages, which contain the capabilities listed in Box 4.2. The Sea Shield mission is to sustain access in contested littorals, to project defensive power from the sea, and to provide maritime defense for the homeland. The Commander of the Third Fleet, together with Commander of the Seventh Fleet, is assigned the responsibility for advancing Sea Shield capabilities. The Third Fleet command ship USS *Coronado* (AGF-11) hosts PACOM's Joint Task Force for Experimentation and acts as the Navy's sea-based battle laboratory to provide a venue for testing new concepts and technology.

The defenses put in place under the Sea Shield pillar envision the use of large numbers of networked, distributed sensors and weapons. FORCEnet obviously has a major role in connecting all of the elements of Sea Shield.

A draft document entitled "Sea Trial Concept Development Plan—Top Level Version 030213," provided to the committee by the Commander of the Third

---

**BOX 4.2**
**Sea Shield Mission Capability Packages**

| Force Protection | Surface Warfare | Undersea Warfare | Theater Air and Missile Defense |
|---|---|---|---|
| • Protect against Special Operations Force and terrorist threats.<br>• Mitigate effects of chemical, biological, radiological, nuclear, and environmental threats. | • Provide self-defense against surface threats.<br>• Conduct offensive operations against surface threats. | • Provide self-defense against subsurface threats.<br>• Neutralize submarine threats in the littorals.<br>• Neutralize open-ocean submarine threats.<br>• Counter minefields from deep to shallow water.<br>• Breach minefields, obstacles, and barriers from very shallow water to the beach exit zone.<br>• Conduct mining operations. | • Provide self-defense against air and missile threats.<br>• Provide maritime air and missile defense.<br>• Provide overland air and missile defense.<br>• Conduct sea-based missile defense. |

Fleet, contains an outline for Sea Shield in the three primary mission areas of Littoral Sea Control, Theater Air and Missile Defense, and Homeland Defense, each further subdivided as shown in Box 4.3. An additional "cut-and-paste" level of detail is provided in the document for each of the bullets given in Box 4.3. No further information on the plans for concept development beyond the list was provided. The lists for each warfare area provide insufficient detail and little indication of what Sea Shield will bring to naval warfare that is new or different in operational concepts from what has been the case historically, although homeland defense is a new area altogether in need of much development.

### 4.3.4.2 Sea Strike

The Sea Strike NCP has four MCPs, which contain the capabilities shown in Box 4.4. The Sea Strike mission is to provide naval power projection focused on offense using both lethal and nonlethal means. The Commander, Second Fleet, in conjunction with the Commander, Fifth Fleet and the Commander, Sixth Fleet, has responsibility for the Sea Strike pillar. Sea Strike capabilities rely on the infrastructure and functionality provided by FORCEnet for command and control and on the ability to transform sensor data and information into actionable knowledge for targeting, maneuver, and strike.

The Second Fleet developed the "Fleet Required Capabilities List for Sea Strike" that includes the following:

- Command and control (C2) and C4ISR interoperability;
- ISR data links to ships;
- Support tools for naval fires;
- Information operations targeting;
- Unmanned vehicles;
- Time-sensitive targeting;
- Jam-resistant technology for weapons guided using the Global Positioning System;
- Tactical decision aids for efficiently and effectively conducting operations including two or more warfare areas at once (e.g., antiair and antisubmarine warfare); and
- Portable and expendable shipboard-launched air targets.

Each of these areas implies the need for some combination of materiel and nonmateriel solution. The Second Fleet has responsibility for identifying and developing nonmateriel approaches where possible, to achieve the required capability. In several of the areas listed the overlap with NETWARCOM responsibilities for FORCEnet is obvious, in that FORCEnet capability will be required to support the fielding of the capability prescribed by the Second Fleet for Sea Strike. Whatever the interaction between the Commander, Second Fleet and

**BOX 4.3**

**Third Fleet Sea Shield Concepts for Sea Trial Concept Development**

**Littoral Sea Control**

1. **Mine Warfare (MIW)**
   a. **Mine Countermeasures (MCM)**
   - Detection, classification, localization, and identification
   - Neutralization
   - Delivery of sensors and prosecutors
   - Network-supported common operational picture
   - Operational and tactical decision aids
   - Refinement of concepts of operations and tactics, techniques, and procedures for MCM and MIW
   b. **Mining**
   - Delivery to selected areas
   - Network-supported command and control and common operational picture including battle damage assessment
   - Operational and tactical decision aids
   - Concept of operations and TTPs for MIW

2. **Antisubmarine Warfare (ASW)**
   - Detection, classification, localization, and identification
   - Neutralization
   - Delivery of sensors and prosecutors
   - Network-supported common operational picture
   - Operational and tactical decision aids
   - Refinement of concepts of operations and TTPs for ASW

3. **Surface Warfare (SUW)**
   - Detection, classification, localization, and identification
   - Neutralization
   - Delivery of sensors and prosecutors
   - Network-supported common operational picture
   - Operational and tactical decision aids
   - Refinement of concepts of operations and TTPs for SUW

**Theater Air and Missile Defense**

1. **Theater Air Defense (TAD)—Air-Breathing Threats**
   - Detection, classification, localization, and identification
   - Neutralization
   - Network command, control, communications, and information
   - Operational and tactical decision aids
   - Refinement of concepts of operations and TTPs for TAD

2. **Theater Missile Defense (TMD)—Ballistic Missile Defense**
   - Detection, classification, localization, and identification
   - Neutralization
   - Network command, control, communications, and information
   - Operational and tactical decision aids
   - Refinement of concepts of operations and TTPs for TMD

**Homeland Defense**

- Combating terrorism/force protection
- Maritime domain awareness
- Predictive analysis
- Maritime interdiction and neutralization
- Refinement of concepts of operations and TTPs

---

**BOX 4.4**
**Sea Strike Mission Capability Packages**

| Strike | Fire Support | Maneuver | Strategic Deterrence |
|---|---|---|---|
| • Conduct strike operations.<br>—Engage fixed land targets.<br>—Engage moving targets.<br>• Conduct special operations.<br>—Provide precision targeting.<br>—Conduct direct action.<br>• Conduct offensive information operations.<br>—Jam potential threats.<br>—Conduct network attack.<br>• Provide aircraft survivability. | • Provide precision fires.<br>• Provide high-volume fires.<br>• Provide extended-range fires. | • Project forces, reposition forces.<br>• Assault centers of gravity and critical vulnerabilities.<br>• Conduct concurrent/ follow-on missions. | • Conduct nuclear strike.<br>• Provide assured survivability. |

---

NETWARCOM, no detail on operational concepts for Sea Strike as enabled by FORCEnet exists. The dual-spiral (coevolutionary) development process may be a useful way to coordinate the nonmateriel solutions brought forward by the fleet with the technological (materiel) solutions that deliver FORCEnet capability.

### 4.3.4.3 Sea Basing

The Sea Basing NCP has three MCPs, which contain the capabilities shown in Box 4.5. The Sea Basing mission is to provide independence, mobility, security, sustainment, and endurance for deployed Expeditionary Maneuver forces. Sea Basing underpins the capability provided through Sea Strike and Sea Shield.

The Second Fleet developed the "Fleet Required Capabilities List for Sea Basing" that includes the following:

- The Joint Force Maritime Component Commander (JFMCC) concept;
- A collaborative information environment;
- A multinational information environment;
- Support tools for intelligence information management, analysis, and fusion;
- Multilevel security;
- Standardized connectivity for coalition information technology;
- Forward logistics;
- Increased mine countermeasure capabilities;

---

**BOX 4.5**
**Sea Basing Mission Capability Packages**

| Deployment and Employment of Naval Forces | Provision of Integrated Joint Logistics | Pre-Positioning of Joint Assets Afloat |
|---|---|---|
| • Close the force and maintain mobility.<br>• Provide at-sea arrival and assembly.<br>• Allow selective offload of materiel.<br>• Reconstitute and regenerate at sea. | • Provide sustainment for operations at sea.<br>• Provide sustainment for operations ashore.<br>• Provide focused logistics.<br>• Provide shipboard and mobile maintenance.<br>• Provide force medical services.<br>• Provide advanced base support. | • Integrate and support joint personnel and equipment.<br>• Provide afloat command-and-control physical infrastructure.<br>• Provide afloat forward staging base capability for joint operations. |

---

- Maritime intercept operations involving noncompliant ships; and
- Support of special operations forces from cruisers and destroyers.

As with the "Fleet Required Capabilities List for Sea Strike," the list for Sea Basing includes several areas that require FORCEnet capabilities for such things as connectivity, information flow, and information management. A draft concept paper for Sea Basing has been written.

#### 4.3.4.4 FORCEnet

The "FORCEnet NCP," which is really a FnII NCP, has three MCPs, which contain the capabilities shown in Box 4.6. This NCP provides the information that enables knowledge-based operations in support of rapid and accurate decision making. Its functionality includes integrating large numbers of diverse, widely dispersed sensors; sharing and processing the sensor data, together with providing the means for delivering relevant information when and where it is needed for decision making; and enabling the means to turn decisions into action.

*Fleet Inputs to FORCEnet.* The fleet's FORCEnet requirements list of the 11 most important items was provided to the committee by the Third Fleet's J9 organization on the committee's visit to the Third Fleet. The list contains these items:

- Data throughput—increase bandwidth;
- Data throughput—dynamic allocation and management tools;
- Migration to Internet Protocol-based C5I (command, control, communications, computers, combat direction, and intelligence);
- 360-degree multiband antenna;
- Multilevel thin client (multilevel security (MLS));
- MLS and collaboration capabilities;
- Real-time collaboration and knowledge management;
- Coalition communications;
- Multiplatform data throughput allocation (ship-to-ship grid management);
- Embarkable and transportable mobile C5I modules; and
- Multi-Tactical Digital Information Link processor.

Most of the FORCEnet requirements listed above are expressed in terms that appear to be preselected technical solutions to operational needs. For example, the 360-degree multiband antenna is a point solution for a reliable (meaning that

---

**BOX 4.6**
**FORCEnet Mission Capability Packages**

| Common Operational and Tactical Picture | Intelligence, Surveillance, and Reconnaissance | Communication and Data Networks |
|---|---|---|
| • Provide mission planning.<br>• Provide battle management synchronization.<br>• Provide common position, navigation, and timing and environmental information.<br>• Integrate and distribute sensor information.<br>• Track and facilitate engagement of time-sensitive targets.<br>• Track and facilitate engagement of non-time-sensitive targets. | • Conduct sensor management and information processing<br>• Detect and identify targets:<br>—Fixed land targets.<br>—Moving land targets.<br>—Air and missile targets.<br>—Surface targets.<br>—Submarine targets.<br>—Mines.<br>• Provide cueing and targeting information.<br>• Assess engagement results. | • Provide communications infrastructure.<br>• Provide network protection.<br>• Provide network synchronization.<br>• Provide information transfer. |

the level of data loss is very low, if any), multiband communications connectivity requirement. Information on how these fleet requirements for FORCEnet relate to and reflect back into each of the Sea Power 21 pillars to provide specific pillar capabilities was not provided to the committee.

***FORCEnet Operational Advisory Group.*** The FORCEnet Operational Advisory Group (OAG), co-chaired by NETWARCOM and MCCDC and composed of members from fleet commands and type commanders, meets twice a year to provide a review and assessment of the needs and requirements of the fleet. The OAG's first meeting, held in July 2003, considered Navy requirements; the second, held in October 2003, examined the needs of the Marine Corps. The top fleet requirements as determined by the FORCEnet OAG are shown in Box 4.7. Rather

---

**BOX 4.7**
**Top Fleet Requirements as Determined by**
**the FORCEnet Operational Advisory Group**

**Requirements**
* Battlespace Awareness
  —Scalable common operational picture (selective availability with confidence)
  —Fused Joint Data Network to and from Joint Planning Network
  —Multi-Tactical Digital Information Link management
* Intelligence, Surveillance, and Reconnaissance
  —Intelligence products to tactical user
  —Fused intelligence connectivity (Blue (Navy) and Green (on-land forces, i.e., Marines and Army))
  —Integrated tasking, processing, exploitation, and dissemination (and user-friendly multi-intelligence analytical functionality)
* Command and Control
  —Component-level command-and-control capability
  —Deployable Combined Joint Task Force/component commander capability
  —Secure Internet Protocol Router chat command-and-control attributes developed (standardized)
* Communications
  —Adequate bandwidth and throughput (50/25 Mb/s)
  —Dynamic bandwidth allocation and management tools
  —360-degree multiband satellite communications connectivity (minimize blockages)
* Focused Logistics
  —Logistics supportability need nodes—afloat and ashore
  —Network access afloat and ashore
  —Addition of medical and maintenance to logistics supportability requirements

---

than using a formal analytic process to relate requirements to warfighting effectiveness, the OAG relies on the collective judgment of the group.

### 4.3.5  Expeditionary Maneuver Warfare and Navy Pillars

Marine Corps Strategy 21 marks the Marine Corps axis of advance into the 21st century. It provides the vision and goals for and aims to support the development of future combat capabilities. EMW is the capstone concept of Marine Corps Strategy 21. It provides the foundation for the way the Marine Corps will conduct operations within the complex environment of the new century. EMW describes the Marine Corps shift from relying on the quantitative characteristics of warfare (mass and volume) to realizing the importance of qualitative factors (speed, stealth, precision, and sustainability).

EMW is firmly aligned with Sea Power 21 and supports the three pillars Sea Strike, Sea Shield, and Sea Basing by providing forces and capabilities directly from the continental United States, from a sustainable sea base, or from adjacent shore locations. The types of interaction that EMW has with each of the pillars is indicated in Box 4.8.

The Naval Operating Concept for Joint Operations describes the naval forces' unique contribution to future joint and multinational operations. EMW capitalizes on congressional tasking for the Marine Corps to operate as an integrated, combined arms force providing a joint force enabler in three dimensions—air, land, and sea. MAGTFs are the Joint Force Commander's optimized force, which will enable the introduction of follow-on forces and the prosecution of further operations.

---

**BOX 4.8**
**Expeditionary Maneuver Warfare Interactions with**
**and Support for Sea Strike, Sea Shield, and Sea Basing**

| Sea Strike | Sea Shield | Sea Basing |
|---|---|---|
| • Persistent intelligence, surveillance, and reconnaissance. | • Sea and littoral superiority forces. | • Maritime Pre-position Forces afloat. |
| • Air or ground forces for time-sensitive strikes. | • Forceable-entry forces. | • Command and control. |
| • Electronic warfare/information operations. | • Theater air and missile defense. | • Offensive and defensive power projection. |
| • Ship-to-objective-maneuver forces. | | • Integrated precision logistics. |
| • Covert strike forces. | | • Accelerated deployment and employment times. |

### 4.3.6 Naval Power 21 Pillars' Feedback into FORCEnet

Considering the Mission Capability Packages contained within the FORCE-net NCP (see Box 4.6) and referring to the capabilities included within the three Sea Power 21 pillars, two points can be made:

• FORCEnet—defined and understood as the network-centric enabler for Sea Power 21—must be considered in the context of the requirements for Sea Shield, Sea Strike, and Sea Basing.
• An implementation strategy for FORCEnet cannot be separated from the implementation of Sea Strike, Sea Shield, and Sea Basing.

All elements of the Navy need to develop a real understanding of the concepts of Sea Power 21 in order to move beyond legacy activities aimed primarily at relabeling old ideas to fit new buzzwords. The pressure to provide something now tends to override thinking about what FORCEnet, as the enabler of Sea Power 21, is supposed to do. The result is that efforts tend to gravitate toward solving old problems rather than toward addressing the new challenges of Sea Shield, Sea Strike, and Sea Basing and the ways in which FORCEnet relates to them to bring real combat capability to those concepts.

At this point, a detailed understanding of what is involved in each of the NCPs is lacking, particularly for FORCEnet. The lists and tables presented above represent most of what is known about the NCPs: that is, several lists and tables have been developed without a great deal of underlying detail. The absence of detail to date can be attributed to the fact that those engaged in this work at the fleet level have other daytime jobs that tend to inhibit making progress in this area. On visits to the commanders of the Second and Third Fleets, the committee observed that only a few people (most on a part-time basis) are engaged in concept development for the three Sea Power 21 pillars. NETWARCOM appeared to have a larger but still inadequate commitment of resources to FORCEnet concept development. Dedicated resources, particularly staff resources, are needed here.

In order to assess and understand the detailed capability needs within each of the Sea Power NCPs, at a minimum the details of the interactions that occur between the Sea Power pillars and FORCEnet at the mission level must be described and evaluated. Such an evaluation should focus on enumerating the dependencies that exist between the Sea Power pillars and FORCEnet and, conversely, the dependencies that FORCEnet may have on the pillars. As discussed earlier, when building to the capabilities envisioned for Sea Power 21, what can or cannot be achieved in FORCEnet depends on what needs to be done operationally within the pillars. Moreover, what can be done operationally within and among the pillars is dependent on the capability that FORCEnet can deliver and when.

## 4.4 THE ROLE OF SEA TRIAL

### 4.4.1 Navy Experimentation

Sea Trial is the Navy process for integrating operational concepts and technology to improve warfighting capabilities through a program of innovation based on experimentation.[8] The Naval Transformational Roadmap of July 2002 puts the fleet in the lead for Sea Trial, and the CNO designated the CFFC as the lead agent for Sea Trial. CFFC Instruction 3900.1A for Sea Trial provides policy guidance, assigns responsibilities, and describes the process.[9]

The Sea Trial concept development and experimentation process is intended to provide a path for rapid maturation of Sea Power 21 concepts and technologies, a way to codify the results of experiments in doctrine, and the means to inform and reflect results in the programs of record. The Sea Trial CD&E campaign plan, maintained by NWDC, is concept-based and aimed at mission capability gaps and fleet priorities. The CFFC Instruction for Sea Trial also establishes the Sea Trial Information Management System (STIMS), in which the details of all Sea Trial experiments are maintained.

The NWDC has been designated the Sea Trial project coordinator and as such is responsible for coordinating the planning and implementation of the Sea Trial process for all the components of Sea Power 21. The Maritime Battle Center (MBC) has been established at the NWDC to serve as the single point of contact for Navy Fleet Battle Experiments (FBEs) and for participation in joint experiments. The MBC is responsible for designing and planning FBEs, as well as for coordinating the execution of these experiments in conjunction with the operational command elements of the numbered fleets and for analyzing and disseminating experiment results. The FBE results are used to accelerate the delivery of innovative warfare capabilities to the fleet, identify concept-based requirements, and evaluate new operational capabilities.[10]

As part of the Sea Trial process, the CFFC has assigned responsibility for prioritization and coordination for the warfighting concept development and experimentation related to each of the Sea Power 21 pillars and FORCEnet to numbered fleet commanders and the commander of NETWARCOM as opera-

---

[8]Another recent study conducted under the auspices of the Naval Studies Board contains additional information on experimentation: National Research Council, 2004, *The Role of Experimentation in Building Future Naval Forces*, The National Academies Press, Washington, D.C. However, that report was not made available to the committee during the course of the present study because it was undergoing Navy classification review prior to its public release.

[9]Commander, Fleet Forces Command (CFFC). 2003. CFFC Instruction, 3900.1A, "Sea Trial," December 22.

[10]Additional information is available at http://www.nwdc.navy.mil/MBC/MBC.aspx. Accessed July 24, 2004.

tional agents. These operational agents "will validate proposed CD&E initiatives . . . oversee the planning, coordination and conduct of Sea Trial events; and brief results. . . ."[11]

The operational agents are to make use of fleet collaborative teams (FCTs) chartered by the CFFC. Generally aligned with the Mission Capability Packages associated with the four NCPs discussed previously, FCTs will provide operational agents with the expertise needed to develop and evolve Sea Power 21 operating concepts.

The CFFC Instruction for Sea Trial is comprehensive. It promotes greater Navy-wide participation in experimentation and the CD&E process. The instruction also establishes greater centralized control over the process for approving experiments, which in the view of the committee could serve to stifle the innovation that experimentation seeks to promote.

### 4.4.2 FORCEnet Innovation and Experimentation at the Naval Network Warfare Command

FORCEnet will be built in an iterative process driven by architecture and experimentation. The process involves the development of a concept and architecture and standards document, subjected to joint and Navy review, that drives the assessment of programs and determination of operational concepts and technical issues that will be resolved through experimentation. This experimentation continuum includes laboratory experiments to evaluate technology, fleet battle experiments to merge operational concepts and systems, and FORCEnet limited objective experiments (LOEs) to test FORCEnet-specific issues and reduce risk for the integrated product demonstrations (IPDs).

NETWARCOM has embarked on a campaign for FORCEnet innovation and experimentation. The approach uses two parallel paths: one for the near term, labeled the FORCEnet prototype path, and another for the midterm and far term, labeled the FORCEnet CD&E path. The prototype path provides the means to field FORCEnet block capability in the fleet immediately to improve joint warfighting. The CD&E path is to feed actionable recommendations from the results of experimentation into the naval capabilities development and planning, programming, budgeting, and execution (PPBE) processes. Both paths look to a process of coevolution among technology, process, and organization. Spiral development is used in a series of fleet-led experiments and exercises that might begin with workshops or war games, spiral into LOEs, and then move to field experiments as a part of fleet exercises. Figure 4.5 illustrates the range of activi-

---

[11]Commander, Fleet Forces Command (CFFC). 2003. CFFC Instruction, 3900.1A, "Sea Trial," December 22.

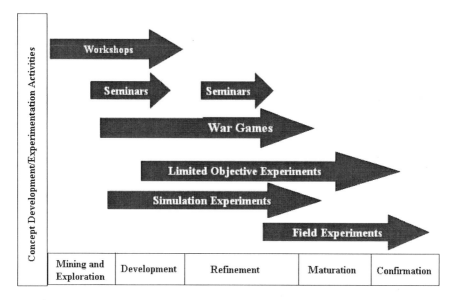

FIGURE 4.5  The range of activities involved in concept development and experimentation for FORCEnet. SOURCE: Commander, Fleet Forces Command (CFFC). 2003. CFCC Instruction 3900.1A, "Sea Trial," December 22.

ties involved in CD&E for FORCEnet.[12] All of these activities together constitute the innovation continuum.

### 4.4.3 FORCEnet Innovation Continuum

Figure 4.6 provides a snapshot of the near-term plan for FORCEnet innovation that includes events ranging from fleet-level experimentation to prototyping to interactions with other joint and Service CD&E in war games. FORCEnet LOEs such as Giant Shadow (LOE 03-1) along the prototype path feed into the bigger events of the annual Trident Warrior series of experiments.

Trident Warrior is established as a series of large-scale Sea Trial events in the joint operational environment. FORCEnet Trident Warrior events are aimed specifically at delivering initial, incremental FORCEnet capability and at developing TTPs and concepts of operations related to the best use of the new FORCE-

---

[12]CAPT Richard Simon, USN, Head, FORCEnet Experimentation and Innovation Group, Naval Network Warfare Command, "FORCEnet Innovation and Experimentation," presentation to the committee on November 17, 2003.

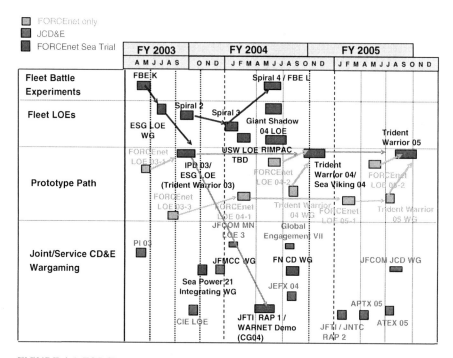

FIGURE 4.6 FORCEnet innovation continuum: near-term plan. (See discussion in text.) NOTE: A list of acronyms is provided in Appendix C. SOURCE: CAPT Roland Mulligan, USN, OPNAV N61F, "FORCEnet—Making Network Centric Warfare a Reality," presentation to the Naval Studies Board, June 11, 2003.

net capability. The focus of Trident Warrior is primarily on developing and experimenting with the FnII.

The first Trident Warrior event—Trident Warrior 03, conducted September 25–30, 2003—was a fleet C4ISR experiment cosponsored by the CNO, NETWARCOM, and SPAWAR to demonstrate FORCEnet capabilities with existing C4ISR products and to deliver the first increment of FORCEnet capability to the fleet. The main focus was on exercising robust, dynamically reconfigurable networks in support of command and control and integrated fires for the ESSEX Expeditionary Strike Group (ESG)[13] in a Pacific Fleet predeployment at-sea exercise in the western Pacific. The demonstrations included bandwidth optimization through allocation, load distribution, and line-of-sight data transfer; distributed and collaborative command-and-control capabilities with a focus on Blue Force Tracking (BFT); and multitiered sensor and weapon information used to generate joint calls for fire.

---

[13] An Expeditionary Strike Group is named after the major ship in the group, in this case, the USS *Essex.*

Network survivability and reliability improvements were realized using four elements—Advanced Digital Network System (ADNS) for dynamic bandwidth allocation, quality of service (QoS) for maximizing load distribution, Challenge Athena III for increased bandwidth, and Inter Battle Group Wireless Network for line-of-sight data transfer. The ability to reroute data transfers in 10 to 30 seconds was demonstrated as links were broken and remade.

Improvements in joint fires operations—reducing target prosecution times from more than 20 to fewer than 5 minutes—were achieved through the use of several technologies associated with the Supporting Arms Coordination Center–Automated.

Following are several SPAWAR observations[14] on Trident Warrior 03:

• The emergent, Web-based information management tools were under-utilized by warfighters owing to the lack of policy fundamentals on how to operate with the tools. This was a noted cultural issue with the forward-deployed naval forces Expeditionary Strike Groups.

• Manpower levels for newly installed fires support systems are deficient. Many systems are only manned by a single person.

• The increased bandwidth that was provided was underutilized.

• Applications were not optimized (Transmission Control Protocol window size) to take advantage of available bandwidth.

Trident Warrior 04 was scheduled for the October/November 2004 timeframe with the TARAWA ESG, to build on Trident Warrior 03 and to provide FORCE-net initial (or Block 1) capability to the fleet. As shown in Figure 4.6, FORCEnet LOE 04-2 and Trident Warrior 04 Wargame precede Trident Warrior 04 and are designed to feed into the larger field experiment. Trident Warrior 04 is a campaign with links to Sea Viking 04, discussed below.

### 4.4.4 Marine Corps Experimentation

The Marine Corps Combat and Force Development process provides for the Marine Corps what Sea Trial provides for the Navy. The Marine Corps is an equal partner in the Sea Trial process. The commanding general, MCCDC, acts as the Marine Corps's Sea Trial coordinator. The Marine Corps conducts the Sea Viking Experimentation Campaign in order to inform decisions and strategies for achieving transformational goals for the year 2015. This CD&E campaign is to assess Marine Corps and Navy capabilities in a joint context. The objective of Sea Viking is to develop and assess the composition and employment of the

---

[14]RDML Michael Sharpe, USN, Vice Commander, Space and Naval Warfare Systems Command, "Architecting and Building FORCEnet," presentation at the FORCEnet 2004 Winning the Information Age meeting, February 10, 2004, San Diego, Calif.

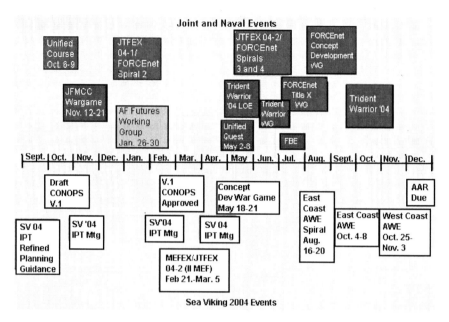

FIGURE 4.7  Major events of Sea Viking 2004. NOTE: A list of acronyms is provided in Appendix C. SOURCE: Marine Corps Warfighting Laboratory, Marine Corps Combat Development Command, Quantico, Va.

future sea-based Marine Expeditionary Brigade, ESG, and Expeditionary Strike Force (ESF) capability sets. A time line of objectives for the Sea Viking is shown in Figure 4.7.

## 4.5  TURNING REQUIREMENTS INTO PROGRAMS

The Sea Trial process diagram (see Figure 4.2) shows two forms of output from operational experiments: "Establish Doctrine" and "System Production." The single arrow connected to each of these outputs represents a number of processes, described below.

### 4.5.1  The Naval Capabilities Development Process

The operational agents for warfighting CD&E related to the NCPs generate requirements for new materiel from the experiments, and the CFFC integrates the operational agents' outputs into fleet capabilities priorities. These priorities are transmitted to the N6/N7 and enter the Naval Capabilities Development Process (NCDP) that validates the requirements and reprioritizes them.

The NCDP deals with Naval Capability Pillars that correspond with the NCPs described in Box 4.1, although the N6/N7 refers to the sets of programs that are intended to supply these capabilities as Naval Capability *Packages*. The "FORCEnet" Naval Capability Package corresponds principally to the FnII and does not span the entire FORCEnet as defined by the CNO. Each Naval Capability Package contains several MCPs that relate to the stated missions each Naval Capability Package is to conduct, and that include specific capabilities that must be realized to some degree in each Naval Capability Package or deployed maritime force package. Note that to the operational agents, the MCPs are sets of operational capabilities; to the NCDP, an MCP is a set of programs that could provide these capabilities.

The NCDP analysis is based on simulations of campaigns set 10 to 15 years in the future. The simulation scenarios assume future strategic situations, enemy threats, and the availability of the materiel already contemplated in previous Program Objective Memorandums (POMs). Experts view the outputs of these simulations, situations that cause unacceptable losses are highlighted, and an attempt is made to identify capability gaps that caused the excessive losses. Excursions are run on the simulations to identify those MCPs that can close the gap so that priority can be assigned to them.

The limitations of the simulation process include the following:

• Setting up a simulation is so time- and resource-consuming that very few scenarios can be examined in a POM process,
• The simulations are of attrition warfare and do not fully represent effects-based operations, and
• Representation of the FnII, surveillance sensors, and C2 nodes and their decision makers are incomplete or inaccurate.

The committee understands that efforts to ameliorate some of the limitations in the third set are contemplated.

### 4.5.2 Resourcing

The N6/N7 directs the Navy's generation and prioritization of internal requirements and passes these priorities on to the N8. The N8 then coordinates the generation of requirements with the Joint Staff through participation in JCIDS.[15] JCIDS, a fairly new construct, requires joint oversight on all DOD acquisition efforts that are anticipated to have lifetime costs in excess of a billion dollars (so called ACAT I programs). This requirements-generation process, lasting up to a

---

[15]For additional detail on the JCIDS process, see Chapter 3 in this report.

year, culminates in a set of joint approved system requirements and priorities being delivered by N8.

The N8 then compares the priorities from the N6/N7 against the Navy's available resources and works closely with the CNO to develop the POM, the DON prioritized budget request submitted to OSD. The ASN(RDA) reviews acquisition program budget changes and new starts during the process to ensure that programs can be certified executable at the programmed levels. Typically, the POM is provided to the DOD about a year before it is anticipated to be voted on by Congress. This year is spent in close discussions with the Congress and the President to develop and pass a final budget.

## 4.6 PROGRAM ACQUISITION

Once Congress has approved funding for a new start, the ASN(RDA) is able to establish a program office and direct the acquisition of the desired technology or capability. (Note that about 2 years have passed since the need was first identified by the fleet.) The ASN(RDA) and staff, and the Deputy Assistant Secretaries of the Navy, typically work through and with the various naval systems commands and program executive offices to design the acquisition strategy needed to meet the need, and to generate the specifications necessary to develop, acquire, and support the needed capability. This process involves not only additional design and feasibility studies, but also work with potential industrial suppliers to ensure that a cost-effective system can be procured. The process requires the development of numerous documents, plans, studies and assessments, mandated by oversight organizations, prior to the release of a request for proposals (RFP). The development of these documents alone can take an additional year or more. One drawback to this stage is that often the offices establishing the original need (the fleet and the DCNOs) are not included in the acquisition stage. This has led to a concern over how the fleet and DCNOs can more effectively ensure that their needs will be met during the final acquisition stage.

As noted, this current process typically takes several years from concept definition to release of an RFP for procurement. The source selection process itself will take another year before a contract can be awarded for beginning to develop the needed capability. Even after no less then six recent acquisition reforms, the time from identified need to contract award is longer than the time to design, build, and deliver a commercial off-the-shelf (COTS)-based capability. This amount of time has been acceptable for large systems (ships, aircraft, spacecraft, and the like), for which basic research or extensive development is required. However, it is not responsive to mission needs that can be satisfied by existing computer-driven technologies in which systems (software and hardware) can go from state of the art to nearly obsolete in the span of a few years.

The NMCI program short-circuited this lengthy process by buying a service versus a system. The time from validation of the need by the Secretary of the

Navy (SECNAV), CNO, and CMC to contract award was less than 1 year. The NMCI program was executed by a PEO empowered by both the acquisition community (ASN(RDA)) and the requirements community (CNO/CMC) to make it happen. The requirement was a one-page document that sufficed because the PEO had representatives on his staff from the requirements community to help develop the procurement specification, and the ultimate users of the capability participated fully in the acquisition and source-selection process. The only impediments hampering speed to capability were oversight organizations external to DON, and even they moved with unprecedented vigor. DOD 5000 series[16] policy changes and perhaps legislation would be required to implement this approach for a product versus a service. Since the complaint about speed to capability is long-standing, perhaps it is time to press for the required system changes and accept the increased risk.

However, the applicability of the NMCI model to FORCEnet is limited. One cannot buy all of the elements that contribute to FORCEnet combat power as a commercial service. Although commercial information technology can make substantial contributions to the FnII, many commercial products assume that the networks on which they ride have continuous high-capacity connectivity—a capability hard to maintain in combat, particularly to ships at sea and dismounted riflemen.

## 4.7 ENABLING TIMELY AND EFFECTIVE COEVOLUTION

The timely and effective coevolution of FORCEnet will require improvements in the three processes shown in Figure 4.1 at the beginning of this chapter (operational concept and requirements development, acquisition and engineering execution, and program formulation and resource allocation) and in the interactions represented by the arrows in that diagram. The material that follows first discusses those activities individually and then discusses the interactions among them.

### 4.7.1 Improving the Activities

The committee observed that the Sea Power 21 operational agents devoted relatively little time to concept formulation. New concepts can inspire the technical community to explore and prototype new capabilities, and an active concept formulation activity would be alert to new technical capabilities and would be devising new concepts to exploit them and experiments to evaluate them. Devoting more resources to concept development would likely be the activity that

---

[16]The 5000 series were the Department of Defense's acquisition process regulations at the time of the Navy-Marine Corps Intranet actions described. These regulations are in the process of being superseded by a capabilities-based process.

would accelerate progress in the CFFC's arena (see Figure 4.1) of operational concept and requirements development.

Some committee members, impressed by the resources of the Pacific Fleet and by the operational insights that the fleet has gained from its mission, believe that this fleet could contribute substantially to concept development and exploitation.

A major limitation of the OPNAV process is the separation of the FnII from the other Naval Capability Packages in the NCDP. This separation causes FnII components to compete for funding with the sensors, weapons, and platforms that they empower. A better approach would be to evaluate the combination of weapons, sensors, platforms, and FnII components that constitute a mission thread. FORCEnet engagement packs could be constructs for such evaluations. The process would be significantly improved if the simulation tool limitations discussed in Section 4.5.1 above were overcome.[17]

All acquisition activity is conducted under the authority of the ASN(RDA). In this capacity, the ASN(RDA) oversees the program executive officers, program managers, systems commands, and ONR. In January 2004, the ASN(RDA) led the first meeting of the FORCEnet Executive Committee (EXCOMM) to address FORCEnet implementation issues. The subjects treated included establishing a FORCEnet implementation baseline and redirecting some current-year funds to support FORCEnet objectives. The decisions and actions of the EXCOMM represent a good start in addressing FORCEnet implementation issues, demonstrating a focus on the future and communicating an urgency for FORCEnet implementation. The meeting, however, had only very limited attendance from the fleet commands; greater senior-level representation of that perspective would aid coordination across all functional areas denoted in Table 4.1 in this chapter.

The committee identified two other opportunities for improving the acquisition process. One, discussed in detail in Chapter 5, would be the promulgation of an architecture that guides functional partitioning and simplifies the integration of new capabilities. The other would be the introduction of some flexibility in the acquisition process.

Today, PEOs presented with specifications and budgets strive for many years to align schedules to meet those specifications, irrespective of what has happened to other acquisitions or to the emergence of new operational constructs and technological opportunities. Keeping the elements of the total FORCEnet synchronized and responding to emerging threats and opportunities requires the ability to change program goals and to reallocate funding among programs without going through the protracted process presently required for major programs.

---

[17]The mission thread analysis planned in conjunction with the FORCEnet baseline assessment could represent a start of the necessary analyses.

Some of the needed flexibility may be attainable by administrative action, but it may be necessary to importune Congress to appropriate a mission thread as a unit and permit the Navy more flexibility in allocating those funds among the components that constitute the thread.

### 4.7.2 Improving Coordination

Beyond improving the three activities designated in Figure 4.1 themselves, there is a need and an opportunity to improve coordination among the activities. As just mentioned, there is no senior fleet representation on the ASN(RDA)'s EXCOMM; that inclusion would improve coordination between the fleet and the acquisition community. More collaborative interaction between the fleet and OPNAV, including the sharing of simulation tools among them, could reduce the dissonance between their respective priority lists. OPNAV should not just pass funded requirements to the acquisition community and wait passively for products to emerge years later. Instead, there should be continuous interaction so that progress in materiel development can be calibrated against changing threats and operational concepts for dealing with them.

### 4.8 GOVERNANCE

Achieving FORCEnet capabilities will require extraordinary process coordination and integration in order to be successful. To oversee this process, the committee believes that a single organization or decision maker may be necessary—one having the mandate and the authority to align inputs across warfighters, technologists, and numerous functional specialists into a coherent, requirements trade-off process. Such an oversight authority would thus need to report to both the CNO on issues of requirements generation, resourcing, concepts of operations, training, and the like; and to the ASN(RDA) on issues related to system prototyping, contracting, procurement, and production.

### 4.8.1 Naval Nuclear Propulsion Model

One model for FORCEnet oversight may be Naval Nuclear Propulsion (NNP). NNP provides a service (nuclear power) to the carrier and submarine communities and so must be responsive to the needs of these communities and must coordinate reactor development time lines to match those of the other ship components. NNP is also responsible for all nuclear power plant concepts of operations, requirements generation, acquisition, R&D, training, and experimentation of naval nuclear reactors. Given NNP's broad roles, its director is given a unique, 8-year term. This long-term awareness and continuous tracking of all relevant activities by the director has been described by many in the Navy as critical to the success and consistency of NNP in producing reliable and useful power plants of the complexity necessary to meet the Navy's needs. Another strength of NNP has been the availability of highly competent technical support organizations.

Although the committee agrees that these two characteristics—leadership continuity and competent technical support—are needed for FORCEnet, it also notes that significant differences exist between FORCEnet and nuclear propulsion. For one, NNP's mandate is over the power plant, while FORCEnet has the potential to impact every individual in the naval Service. NNP was able to start from scratch (no nuclear reactors existed in the Navy before the office was created), whereas any FORCEnet oversight authority will have to spend significant effort aligning legacy systems and existing initiatives.

### 4.8.2 Future Combat System Model

The Army's FCS is comparable in scope to FORCEnet in that its materiel aspects include platforms, weapons, and sensors, as well as the equipment that networks them. The Army's approach to the FCS was to compile a detailed performance specification and then to select a systems integration contractor to which it granted authority exceeding that usually granted to a prime system contractor. The integration contractor decomposes the performance specification into systems, acquires them, and, after they are delivered, will integrate them into the FCS.

The committee expresses little enthusiasm for applying the FCS model to FORCEnet. The FCS model presumes a fixed end state that can be described by a system specification, and it appears to make little provision for the coevolution of concepts and materiel capabilities.

### 4.8.3 Programs-Acquisitions Coordination Board

Recognizing that authority over acquisitions is vested in the ASN(RDA) and that authority over programs and resources is vested in OPNAV, another option for managing process coordination and integration calls for the formation of a Programs-Acquisitions Coordination Board, co-chaired by the Vice Chief of Naval Operations (VCNO) and the ASN(RDA) to synchronize and coordinate FORCEnet activities in both their domains.[18] Because of the broad scope of FORCEnet, the scope of this board would be tantamount to that of a General Board.

The Programs-Acquisitions Coordination Board would have support from a dedicated staff in OPNAV and the office of the ASN(RDA) to monitor events in

---

[18]Including the Commander, Fleet Forces Command, or Commander, Naval Network Warfare Command, on the board could resolve the differences between the requirements priorities of the Office of the Chief of Naval Operations and the fleet.

both domains and to present issues to the board. The board would meet regularly, and the staff would work issues on a daily basis. The executive secretary of the board, a flag officer or senior executive service equivalent, should have tenure longer than the 2 to 3 years typical for flag assignments.

The Naval Studies Board committee that 5 years ago considered the challenges of realizing network-centric capabilities made a similar recommendation in its report.[19]

### 4.8.4 Director of FORCEnet

Although most of the committee believes the Programs-Acquisitions Coordination Board to be superior to the two previous options—the Naval Nuclear Propulsion or the FCS model—some are pessimistic about the forcefulness of a board. They would prefer to give the responsibility for synchronizing and coordinating all of these activities to a single individual, the director of FORCEnet, an O-9 or O-10 who would serve longer than the typical 2- or 3-year term. Because of Goldwater-Nichols requirements, the director of FORCEnet would report to the ASN(RDA) for acquisitions matters and to the CNO or VCNO for other matters. The director of FORCEnet would be supported by the same staff that the Programs-Acquisitions Coordination Board would have.

Because all PEOs must by law report directly to the ASN(RDA), DON's senior acquisition executive, the director of FORCEnet, would not have line authority over the PEOs. However, if the ASN(RDA) followed the advice of the director of FORCEnet, or if the CNO/VCNO gave the director of FORCEnet control of the funds on which the PEOs depend, the director would have sufficient authority over the PEOs[20] to assure that all materiel was acquired in conformance with the FORCEnet architecture. However, giving the director of FORCEnet such wide control over funds might seem to be usurping the authority of the N6/N7 and N8.

### 4.8.5 Synthesis

The committee agrees that a mechanism is needed to synchronize OPNAV's program responsibilities and the ASN(RDA)'s acquisitions responsibilities. It

---

[19]Naval Studies Board, National Research Council. 2000. *Network-Centric Naval Forces: A Transition Strategy for Enhancing Operational Capabilities*, National Academy Press, Washington, D.C., p. 7.

[20]Some committee members suggest that the problem could be largely solved by merging the PEOs for Integrated Warfare Systems, for Information Technology, and for Command, Control, Communications, and Information into a "super-PEO" led by the director of FORCEnet. There would remain the need to give the director sufficient authority over the PEO for Submarines, PEO for Air, and so on, in order to assure their conformity to the FORCEnet architecture.

recommends either the creation of the Programs-Acquisitions Coordination Board or the appointment of a senior director of FORCEnet reporting both to the CNO or VCNO and to the ASN(RDA). However, the committee does not have unanimity as to the relative desirability of these two options. Some members believe that boards are not effective and that a strong director of FORCEnet is the only workable option. Others believe that the options are equivalent, because full empowerment of the director would be impractical, and he or she would be relying on the authority of the VCNO and the ASN(RDA), who would be the co-chairs of the board if there were one.

All members of the committee agree that the strength and continuity of the director of FORCEnet or the chief of staff of the Programs-Acquisitions Coordination Board are essential factors for success in implementing FORCEnet, as is the quality of the staff support to the director of FORCEnet or the Programs-Acquisitions Coordination Board.

### 4.8.6 Oversight by the Chief of Naval Operations

Whatever governance mechanism emerges to coordinate and integrate FORCEnet-related activities, means are needed to keep a fleet perspective in monitoring and accelerating the deployment of new capabilities. Logically, the CFFC would be responsible for reporting to the CNO both on the development of operational constructs and on the effectiveness of materiel being deployed to support these constructs. By periodically setting goals for new operational capabilities, the CNO could provide oversight to the development of both constructs and technical capabilities. A CNO-driven, annually revised master plan with goals stated in terms of operational capabilities to be realized in the near term would motivate all parties.

## 4.9 FINDINGS AND RECOMMENDATIONS

### 4.9.1 Findings

Following are the committee's findings with respect to the three major FORCEnet implementation activities—operational concept and requirements development, program formulation and resource allocation, and acquisition and engineering execution—and the prospects for improving their coordination.

Coevolution using the dual-spiral approach for the development of the operational construct and architecture of FORCEnet provides positive opportunities for interaction between operators and acquirers as a means to validate solutions for FORCEnet capability needs and gaps. Coevolving technology with concepts, doctrine, and other nonmateriel solutions through greater interaction among users and developers can speed the delivery of improvements in warfighting capabilities to the fleet.

However, very little detail has been developed articulating new operational concepts—only limited descriptive material and certainly nothing with the sort of detail typically found in operational architectures.[21] This fact is most likely a consequence of the very limited resources committed to this area. The Second and Third Fleets devote only a few people part time to concept development for the three Sea Power 21 pillars. NETWARCOM appears to have a larger, although still small, commitment of resources to FORCEnet concept development. Interaction between the pillars and FORCEnet as the enabler is very limited. Representatives of the organizations mentioned, especially the Second and Third Fleets, indicated that these limited commitments were a consequence of the many demands (e.g., maintaining readiness) placed on these organizations.

All organizations indicated a serious commitment to experimentation, although generally one of modest scope. The Second Fleet has been active in exploring the use of prototype equipment, the Third Fleet has a history of experimentation centered on the USS *Coronado* command ship, and NETWARCOM is conducting the Trident Warrior series of exercises, with its experimentation thus focusing largely on the FnII.

The CFFC has underscored the importance of experimentation by issuing a new experimentation instruction (CFFC Instruction 3900.1A for Sea Trial). Furthermore, the CFFC reduced the number of the large fleet battle experiments to allow more of the smaller, limited objective experiments, which should promote greater exploration and innovation. The Sea Trial instruction promotes greater Navy-wide interaction, thereby potentially bringing more ideas and resources to experiments. At the same time, though, this instruction establishes greater centralized control in approving experiments, which could stifle the very innovation that experimentation seeks to promote.

NETWARCOM has an active program for the development of FORCEnet requirements, drawing widespread community participation though its Operational Advisory Group. It does not use any formal analytical methods to relate the requirements to warfighting effectiveness, relying rather on the collective judgment of the group.

The Second and Third Fleets demonstrate only very limited requirements development for the three Sea Power 21 pillars of Sea Shield, Sea Strike, and Sea Basing. While fleet FORCEnet requirements lists have been made, very little interaction of the three pillars with FORCEnet is evident. This limited work is most likely a consequence of limited resources, as described above for operational concepts development.

The NCDP used by OPNAV for formulating and prioritizing programs in response to fleet requirements has not fully explored the interactions between the

---

[21]In March 2004, after the cutoff date for new input to this study, the Naval Network Warfare Command initiated an effort to develop a FORCEnet operational concept, which could provide a more detailed product.

FnII and other FORCEnet components and may have led to a competition be-tween the FnII and the other components that it empowers. The modeling and simulation tools used in program assessment are less than ideal, although defi-ciencies are recognized and efforts to ameliorate some of them are contemplated.

Resourcing of programs takes place years after the need for them has been recognized, and program managers have insufficient flexibility to respond quickly to changes in threats or to new concepts or technological opportunities. The need for more flexibility is particularly acute with respect to the FnII because of the rapid pace of technology advancement.

The influence of the fleet and OPNAV is greatly diminished once a program enters acquisition. If the coevolution of FORCEnet concepts and technology is to be effective, tighter coupling is needed among the activities depicted in Figure 4.1, and mechanisms are needed to accelerate speed to capability.

### 4.9.2 Recommendations

Based on the findings presented above and on the issues described in this chapter, the committee recommends the following:

• **Recommendation** for NETWARCOM, and the Second and Third Fleets especially: Devote significantly more resources to concept development. The criticality of concept development to the overall realization of FORCEnet capa-bilities certainly requires this increase. The committee recommends that CFFC determine whether the increased resources would come by reassigning personnel already assigned to the organizations or by request to the CNO for additional personnel.

• **Recommendation** for the CNO: Assign the Pacific Fleet greater direct responsibility in Sea Power 21 concept development. This action would apply the sizable resources and operational experience of Pacific Fleet to help redress the current limitations in resources devoted to concept development. The action would also help strengthen the joint aspects of concept development through Pacific Fleet's relation with PACOM.

• **Recommendation** for CFFC: Ensure that NETWARCOM plays as broad a role in FORCEnet concept development and experimentation as possible—not just limited to the use of the FnII. This is consistent with NETWARCOM's charter and reflects the fact that FORCEnet involves forcewide capabilities.

• **Recommendation** for CFFC: Ensure that the centralized management processes of the new Sea Trial instruction do not stifle innovation. Local initia-tive is critical to innovation. The Sea Trial management mechanisms should concern themselves with setting broad guidelines and resource allocations within which individual elements in the Navy would be free to innovate. Every experi-ment, no matter how small, should not require approval by a centralized commit-tee, as would appear to be the case with the new Sea Trial instruction.

• **Recommendation** for NETWARCOM: Develop analytical means for the development and prioritization of requirements. This would allow requirements to be tied better to warfighting effectiveness and would thereby better support these requirements in the resource-allocation process.

• **Recommendation** for the Second and Third Fleets: Devote more resources to the development of requirements for the three Sea Power 21 pillars. Needed capabilities for the pillars must be adequately specified in order to determine the necessary FORCEnet capabilities. Means to obtain these resources would be addressed by reassigning personnel already assigned to the organizations or by request to the CNO for additional support.

• **Recommendation** for the N6/N7 and N8: Develop resource-allocation methods directed at realizing forcewide FORCEnet capabilities. Instead of basing the methods on the current Naval Capability Packages, the Navy should instead use "packages" that inherently reflect network-centric operational concepts. FORCEnet Engagement Packs provide one such example.

• **Recommendation** for the N6/N7 and N8: Develop (or acquire) modeling and simulation tools that allow faster exploration of scenarios and better measurement of the effects and limitations of information availability and network connectivity in warfare. This will not be an easy task since such tools are in their infancy, but the Navy should be a proponent for the development of these tools.

• **Recommendation** for the ASN(RDA): Take action to include senior members of the fleet commands in the deliberations of the FORCEnet EXCOMM. Their perspective in general would be useful. In particular, the actions necessary to implement FORCEnet capabilities in a fixed-resource environment could impact near-term fleet readiness and should be accomplished in partnership with fleet representatives.

• **Recommendation** for the ASN(RDA): Explore methods for increasing flexibility in resource allocation. One approach for doing so is to aggregate program line items into larger line items, including the possibility of establishing a few major lines referring to FORCEnet capabilities (e.g., for implementation of the FnII or for the systems engineering required across the entire fleet). The Navy, in conjunction with the other military Services, could also consider approaching Congress to relax the limit on reallocating program funds. A strong argument for this authority could be made on the basis of the current need to field systems of systems, in contrast to the previous focus on individual systems.

• **Recommendation** for the ASN(RDA): Review Navy acquisition processes and practices and institute educational measures as necessary, to ensure that programs are providing as rapid a delivery of capability as possible. For example, financial practices could be reviewed to determine means for emphasizing rapid capability delivery while maintaining accountability, and execution instructions could be reviewed to ensure that there is adequate delegation of authority.

- **Recommendation** for the SECNAV, in conjunction with the CNO and the ASN(RDA): Develop a means to integrate more closely the Navy's program-formulation and acquisition functions, to ensure that adjustments in program execution are consistent with program intent and best serve the overall need of providing forcewide FORCEnet capability. Options to consider include establishing (1) a Programs-Acquisitions Coordination Board co-chaired by the VCNO and the ASN(RDA) or (2) a director of FORCEnet reporting to the VCNO and ASN(RDA). This recommendation envisions that the board or director (depending on which was chosen) would have a major role in carrying out the other recommendations pertaining to program formulation and resource allocation and to acquisition and engineering execution.

- **Recommendation** for the CNO: Charter the CFFC to provide periodic assessments of the state of realizing FORCEnet capabilities. The review would include the following: the status and plans for concept development and experimentation for each of the Sea Power 21 pillars and FORCEnet, the current understanding of the set of capabilities required in the fleet, recommended changes in programs to align them better with this set of capabilities, and opportunities for employing acquisition prototypes in naval and joint experiments and exercises. NETWARCOM would provide the staff support to the CFFC in preparing this assessment.

- **Recommendation** for the CNO, in conjunction with the ASN(RDA): Establish a set of FORCEnet goals to be realized by specified dates in order to drive the implementation process. Examples of these goals include the provision of specified bandwidth increases and networking capabilities to the fleet, the achievement of designated joint maritime and air situational-awareness capabilities, and the achievement of FORCEnet compliance (or phaseout) for a specified set of legacy systems. Goals could also be of a directly operational nature—for example, the ability to destroy a given class of targets within a stated number of minutes after the targets emerge from hiding.

- **Recommendation** for the CNO, in conjunction with the ASN(RDA): Direct the preparation of an annual FORCEnet master plan for their review. The plan should lay out milestones—with an emphasis on near-term deliverables—for obtaining key FORCEnet capabilities in terms of operational concepts and systems deployment. The purpose of this plan would be to ensure senior visibility and scrutiny of FORCEnet activities and consequent motivation for conducting these activities.

# 5

# FORCEnet Architecture
# and System Design

If FORCEnet is to be the architectural framework for naval warfare in the information age, it must deliver performance, information assurance, and quality-of-service guarantees unprecedented in a system with the nodal diversity evidenced in the joint force. This challenge is best met incrementally so that existing capability is not degraded nor information security ever compromised. The design and implementation of complex systems for purposes of warfighting require a dedicated core of warfighters and system engineers trained in the art of operations analysis. Together, warfighters and engineers make decisions about when and how to introduce new capabilities as technologies and operational concepts evolve in independent but integrated spirals, as described in Chapter 2. Traditional, rigid system engineering cannot accommodate the technology cycle as it applies to complex IT-enabled systems. However, the fundamental principles of system engineering still apply. This chapter addresses the following topics: technical characteristics that are required to achieve the full realization of FORCEnet, architecture definition and process, the status of efforts to realize the FORCEnet architecture, system engineering principles and considerations, managing operations, facilities, and technical dangers and solutions.

## 5.1 WHAT IT TAKES TO ACHIEVE THE FORCEnet PROMISE

### 5.1.1 Guaranteed End-to-End Quality of Service

It is imperative to recognize that a network-centric force is a network-dependent force. Network dependency demands access and quality-of-service guarantees that can be adjusted to reflect mission and COI priorities. Suggested metrics

*115*

for quality of service are addressed elsewhere (see Table 3.3, Table 6.3, and "Network Quality of Service and Resource Management in a Military Context" under Section 6.2.1.1), but whatever the final set, all network warfighting nodes must have the capability to know the state of health of the force combat system of which they are a part.

### 5.1.2 Bandwidth

Guaranteed quality of service requires, among other things, guaranteed bandwidth availability. Experience has shown that availability of bandwidth has been problematic in the naval environment, and that is not expected to change in the near future. The Congressional Budget Office and the U.S. Army have estimated military bandwidth demand at up to 20 times greater than forecast capacity.[1] DISA estimated that more than 80 percent of the bandwidth consumed during OIF was supplied by commercial sources, and even then there were bandwidth brownouts.[2]

Current plans for bandwidth expansion to naval forces do not adequately address the requirements of a mobile network-centric force prior to the launch of the TSAT. Initial TSAT operational capability is scheduled for 2012 or later, which seems out of sync with plans for near-term initial releases of horizontal data fusion and enterprise services as envisioned by the ASD(NII). Horizontal data fusion and enterprise services will further increase the demand for bandwidth. For example, Extensible Markup Language (XML) encoding of data can easily increase message size by a factor of 10 or more, and it has been estimated that Littoral Combat Ships (LCSs) will each require bandwidth of 8 megabits per second (Mb/s) continuous and 20 Mb/s burst when they are deployed. Where will this bandwidth come from? What happens to the Web services environment and the LCS investment if it is not available?

The committee was not able to uncover good estimates of the bandwidth and quality of service required for the envisioned networked services. NMCI leveled the QoS playing field between disadvantaged and advantaged users ashore. A similar effort is required to level the quality of service playing field between fixed and mobile users that project power to the edge. Since there will likely never be enough bandwidth to satisfy all needs simultaneously, agile bandwidth allocation and distribution schemes are necessary, as is disciplined design control of bandwidth demand. Since the LCS will likely be the first ship class that is highly dependent on external networks and resources to accomplish its mission, it is an excellent candidate for the application of bandwidth-demand control techniques.

---

[1]Congressional Budget Office. 2003. *The Army's Bandwidth Bottleneck*, U.S. Government Printing Office, Washington, D.C., August.

[2]"DISA Chief Outlines Wartime Success," *Federal Computer Week*, June 6, 2003.

### 5.1.3 Information Assurance

The term "interdependence" requires high levels of system and operator maturity as well as an information assurance architecture and concepts of operations that are agile and robust enough to respond to emerging information assurance threats, while at the same time supporting information sharing. A risk–benefit analysis should be conducted before a decision is made to field a design that makes a fighting unit dependent on assets that it does not control. It is not yet clearly understood what can go wrong or how to assess FORCEnet value in light of such risks, but it is known that denial-of-service attacks are real and must be considered in the risk analysis. Since there is no known 100-percent-effective defense against denial-of-service attacks, graceful degradation from connected to independent operations must be a critical parameter in the design of the information assurance architecture and concepts of operations. The FORCEnet security architecture work that is in progress is commendable and must continue with high priority throughout the design and implementation process. The security architecture must permeate FORCEnet end to end, from network access controls through applications access to data access within applications.

### 5.1.4 Availability, Redundancy, and Graceful Degradation

In order to ensure quality of service in FORCEnet, there must be much higher levels of availability of network assets than have been achieved in the past. This change will require the addition of redundancy and graceful degradation, including automatic network reconfiguration, for occasions when failures occur. FORCEnet must be built with network management in mind. The warfighting value of FORCEnet capabilities and the speed of implementation will be paced by bandwidth availability.

An example of requisite availability is "multi-nines" for critical functions, such as weapon midcourse control, for which only a few hours or minutes per year are tolerated. A more typical wireless availability of 90 percent would generally imply that connectivity is down about 36 days per year, generally for short intervals, and that this level of availability is adequate for the networked functions. To achieve even the case of a 90 percent available design requires accommodation for propagation fading and rerouting.

Distributing system functionality for systems that were not originally designed to be part of a distributed system generally reduces the overall availability of the functions performed, unless redundancy is introduced. This is simply because there are more mission-critical parts in the system. Increased capability, however, can offset lower availability and increase the probability of mission success. The probability of mission success for a system can be thought of in terms of three independent variables: the probability that the system is available times the probability that the system has the capability to perform the mission times the probability that the operators will perform without error. The probability of mission success cannot be higher than the lowest of any of these variables. Capability increase through the networking of assets is frequently a high-return-on-investment option for increasing the probability of mission success.

### 5.1.5 Architecture That Supports Incremental Deployment

Inevitably it will be too expensive to deploy new capability to every fighting unit simultaneously. Thus, there will always be a mix of old and new that must work together without degrading legacy performance. These realities will likely lead to the need for a transition architecture that may be missing desired characteristics and capabilities, but which will start the process of getting capability into the fleet. For example, a hub-and-spoke configuration with ad hoc networking could support near-term capability enhancements.

Multiple hubs supported by an airborne relay could provide needed redundancy, while allowing limited satellite bandwidth to be shared between assets within a 200 mile radius of the relay. This configuration could also support incremental deployment, since a node that is part of two subnetworks of varying link configurations can be designated as the communications filter between old and new configurations.[3] (Mobile ad hoc networks are discussed in Chapter 6.)

### 5.1.6 Interoperability

The Navy and DOD have worked for 12 years to develop a single integrated air picture and have yet to do so. Many factors contribute to interoperability and to the lack thereof. One factor, however, seems clear: absent the market pressure of the commercial marketplace, the complexity of the problem of interoperability is greater than can be solved by the enforcement of standards. To illustrate the complexity, assume that one wants to fuse data from multiple sensors. If data from multiple sensors are to be fused and not just presented in their native form, all of the sensors must "know" where they are relative to all other sensors, must know what time it is, and must know with what precision they are reporting location, time, and the position. Since most of today's sensors, identification algorithms, track correlation algorithms, databases, time standards, and so on were not designed with the idea of fusion in mind, the statistical confidence in the reported data is not known, and thus the statistical confidence in the fused data is unknowable. FORCEnet must establish and implement standards and information architectures that result in deterministic outcomes when data are fused. These

---

[3]This approach is illustrated in the paper "Battle Force Interoperability and Net-Centricity," by Joseph Cipriano, Naval Sea Systems Command, and Jerry Krill, Johns Hopkins University Applied Physics Laboratory, 1999, *Defense Systems and Equipment International*, Laurel, Md., pp. 261-270.

go far beyond communications standards. The proper allocation of functions, as discussed in Section 6.2, will facilitate the process, but the variables determining interoperability are so numerous that it should not be expected that these issues will be corrected by planned FORCEnet activities. Matters such as processor speeds and buffer size, in addition to the more obvious transport-layer and data-definition issues, will present additional challenges. A common transport reduces the number of interfaces that have to be maintained from an $O(n^2)$ problem to an $O(n)$ problem, where $n$ is the number of systems that are to interoperate. Functional partitioning that results in common boundaries for common functions and common data definitions reduces the data fusion complexity from $O(n^3)$ to something more manageable.

## 5.2 ARCHITECTURE DEFINITION AND PROCESS

### 5.2.1 Architecture Defined

The FORCEnet information architecture should be thought of as a boundary between layers of functionality that is held invariant (over long periods of time), thus allowing developments to proceed independently on all sides of the boundary. In the committee's view, architecting FORCEnet is the process of defining thin waists, or boundaries, that are invariant and, when coupled with selected industrial standards and throttled with a network control system, would enable FORCEnet to evolve with advances in technology. The boundaries standardize the interfaces between the functions common to all warfare systems so as to facilitate interoperability and information sharing. Examples of boundaries that should be established include those between sensor/intelligence networks, command-and-control networks, fire-control networks, displays, and databases, as shown in Figure 5.1. The open-architecture initiative of the PEO for Integrated Warfare Systems (IWS), discussed later in this chapter, is an application of this concept at a lower level of system detail.

The establishment of the boundaries shown in Figure 5.1, populated with designs that are compliant with the Networks and Information Integration (NII) Net-Centric Checklist and validated with performance modeling and sensitivity analysis, would provide a solid foundation for the FORCEnet information architecture. The network control system could then be designed to route and throttle information across the boundaries on the basis of mission priorities and available capacity and capability. As envisioned, any sensor could be accessed by any command-and-control system, which in turn can direct any fire-control system through a human in the loop. Multiple echelons are reflected in communities of interest related by the common boundaries that they share.

There are several areas in which boundaries can be defined in a service-oriented architecture, but with too many boundaries there is insufficient homogeneity to operate as an enterprise. In the commercial marketplace, market pressure

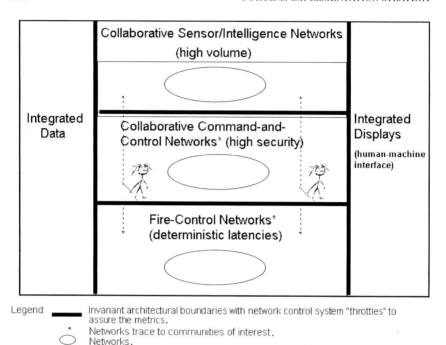

Legend:  ▬▬▬▬  Invariant architectural boundaries with network control system "throttles" to
                assure the metrics.
          *     Networks trace to communities of interest.
          ◯     Networks.
        ------  Kill chains.

FIGURE 5.1 Examples of boundaries between functions in the FORCEnet information
architecture.

limits the number of boundaries. In the DOD, governance must be relied on to
limit them.

### 5.2.2 Architectural Process

As briefed to the committee, several groups felt that they were in control of
the architecture definition process. "Generate-and-test" is a costly trial-and-error
approach. Architectures are best defined by a small group of bright, experienced
designers. Architectural principles can be defined and applied in a systematic
manner to create architectural integrity. A prime example is the development of
the Internet. Commencing with the vision for a communications infrastructure
that could survive attacks by adversaries, packet switching (breaking data into
packets for independent routing) became the architectural principle. Ten years
later, IP became the architectural definition.

An example of an effective process is one called user-centered design, in
which desired functionality is defined in terms of "use cases." The Navy sometimes calls these use cases Design Reference Missions, and the joint community
calls them mission threads. Capability requirements are derived from the use

cases, and the architecture is synthesized to provide these capabilities. Design and implementation experts select technologies and standards to realize the architecture. Components are created to support unique requirements. Glue logic, which can take the form of middleware or concepts of operations, is synthesized to tie together existing technology that is not interoperable until the maintenance burden of the glue becomes greater than the value of the interoperability. When the maintenance burden becomes high, functional boundaries need to be modified to reduce interoperability issues and cost. If the boundaries are set correctly, this should be an infrequent event. Since the existing functional boundaries were set when the definition of system did not extend beyond a warfighting platform, many functional boundaries are in need of redefinition. By analogy, when every home had a well, the interface between the water supply and the kitchen was a bucket; the bucket could still be used when a public water supply became available, but why would one do so?

It appears to the committee that the Air Force Command and Control, Intelligence, Surveillance, and Reconnaissance Center (AF C2ISR) has the closest to a user-centered design process of any of the Services. The threads of the AF C2ISR are examples of use cases. In order to be robust and provide room for growth, more visionary use cases and mission threads should be defined for FORCEnet prior to architecture synthesis.

## 5.3 STATUS OF EFFORTS TO REALIZE
## THE FORCENET ARCHITECTURE

### 5.3.1 Architecture Documents of the Space and
### Naval Warfare Systems Command

SPAWAR is developing a series of reports to help define the FORCEnet architecture. The Version 1.1 release of November 18, 2003, consists of the first two volumes:

• *Volume I—Operational and Systems View* describes FORCEnet as a standards-based open architecture.[4] Volume I consists primarily of a high-level survey of military systems that would form components upon which FORCEnet will be realized. The output of the FORCEnet Baseline Process, briefed to the FORCEnet EXCOMM on June 23, 2004, will undoubtedly modify the list as systems are placed in compliance categories.

---

[4]Space and Naval Warfare Systems Command. 2004. *FORCEnet Architecture and Standards, Volume I, Operational and Systems View*, Version 1.4, San Diego, Calif., April 30. The version of the SPAWAR architecture reviewed by the committee was Version 1.1, November 18, 2003. A brief review of the later Version 1.4, April 30, 2004, was also made and did not affect the committee's conclusions.

• *Volume II—Technical View* is a list of almost 300 mandated standards that FORCEnet compliant systems must support.[5]

Neither Volume I nor Volume II appears to define the Navy-specific or joint invariants or boundaries in the FORCEnet architecture. Invariants allow technology to evolve on all sides of architectural boundaries, sometimes by orders of magnitude, and still maintain functionality. Following is a summary of the contents of these two volumes, as well as suggestions on Navy-specific invariants that could be defined.

### 5.3.1.1 Summary of *FORCEnet Architecture and Standards: Volume I, Operational and Systems View*

The preface to Volume I states that "the FORCEnet Architecture will incorporate common engineering, information, protocols, computing, and interface standards . . ." and that "the FORCEnet Architecture is based on a commercial Distributed Services model." Further, FORCEnet is called a "standards based open architecture."[6]

As listed in Volume I, the following FORCEnet architectural principles have been adopted:

• Standard, published interfaces;
• Separation of interface from implementation;
• Open architecture;
• Task-centered design at the presentation layer;
• Database independence;
• Joint interoperability;
• Uniformity in architecture and design; and
• Recognition of diversity.

Volume I indicates that six core architectural elements have been identified: ISR (intelligence, surveillance, and reconnaissance) systems; weapon systems; command-and-control and support systems; network services; networks; and communications systems. In addition, a dozen IT service categories have been defined: networking, identity management, security, operating system, user (person)-to-computer interface, data management, data interchange, multimedia/

---

[5]Space and Naval Warfare Systems Command. 2004. *FORCEnet Architecture and Standards, Volume II, Technical View*, Version 1.4, San Diego, Calif., April 30.

[6]Space and Naval Warfare Systems Command. 2004. *FORCEnet Architecture and Standards, Volume I, Operational and Systems View*, Version 1.4, San Diego, Calif., April 30, pp. 2-11.

graphics, communications, document management, support, and hardware. Service components are identified for each of these service categories. A service component is made up of standards, interfaces, protocols, and product specifications. Examples of possible service components by service category include the following:

- *Networking:* network management, address management, and routing protocols;
- *Identity management:* federated directory service;
- *Security:* identification, authentication, audit trail creation and analysis, access controls, cryptography management, virus protection, intrusion prevention and detection;
- *Operating system:* kernel operations, fault management, utilities, backup and recovery;
- *User (person)-to-computer interface:* dialogue support, window management, and multimedia; and
- *Data management:* metadata, data dictionary, directory services, database management system.

The FORCEnet functional architecture is based on MCPs. These are not the MCPs that the Warfare Integration Unit under the DCNO for Warfare Requirements and Programs (N70) uses for program assessment. Instead, two operationally oriented scenarios have been defined to validate the FORCEnet architecture: (1) time-critical targeting employing persistent sensors and (2) cruise missile defense.

The approach in *Volume I, Operational and Systems View*, is to list current and emerging programs that either will form the FORCEnet infrastructure or will have to interface with the infrastructure. The architecture builds on existing systems and proposes upgrade roadmaps in the following areas:

- *Communications and networks:* migration and consolidation from existing systems and their planned extensions, including NMCI (land-based); NCTAMSs/Navy computer and telecommunications stations (ships at pierside and underway); FLEETnet global IP routing; tactical data links; and radio;
- *ISR:* DOD DCGS; GIG-enterprise services (GIG-ES) (access, analysis, storage, dissemination); joint C2 (warehouse of information, including force tracking, intelligence, maps, weather, socioeconomic, and cultural). Data will be in an XML infrastructure that manages data types by using wrappers and encryption for each data object; and
- *Distributed services:* based on Service-oriented architecture with registration, discovery, and Service interface; dozens of networking standards are listed. The Open Architecture Computing Environment (OACE) is the basis for services in FORCEnet. The GIG-ES could supply many of the basic services.

## 5.3.1.2  Critique of Volume I

Volume I includes numerous systems that are deployed and/or under development. This committee finds it difficult to understand how the pieces fit together and what is architecturally tying them together. One possible description of FORCEnet is that it is a system of systems, but the interactions between these systems may be more opportunistic than systematic. Definition of invariant boundaries as suggested in Section 5.2.1 would greatly enhance the value of both volumes.

## 5.3.1.3  Summary of *FORCEnet Architecture and Standards: Volume II, Technical View*

Volume II focuses on information technology standards (as opposed to capabilities). Standards were selected on the basis of their supporting interoperability and interchangeability, their maturity, ease of implementation, public availability, and consistency with public law and regulations. The FORCEnet technical working group will review new standards proposed for inclusion. In addition, a FORCEnet Compliance Process and Checklist is under development.

Almost 300 standards are mandated in Volume II. They cover such areas as data formats; operating system services; sensor interfaces; position, navigation, and time; information assurance; and information transfer. In areas in which standards are mature, they are mandated. For example, some of the mandated standards in the area of data formats include the following:

- Databases (Structured Query Language);
- Documents (Standard Generalized Markup Language, Hypertext and Extensible Markup Languages);
  - Geospatial information (raster, vector, maps);
  - Audio;
  - Atmospheric and oceanographic data;
  - Time of day;
  - Floating point numbers; and
  - Graphs.

## 5.3.1.4  Critique of Volume II

Volume II is an extensive list of standards that FORCEnet-compliant systems must support. The list includes required standards and emerging standards that might migrate to the required list. As standards evolve, the list becomes a moving target, almost guaranteeing interoperability problems.

The process of using standards to achieve interoperability has not always been successful in the past in the DOD. In the commercial arena, standards achieve interoperability through market pressure that does not exist for DOD-unique standards. Any deviation from widely used commercial standards will likely yield unsatisfactory results.

### 5.3.2 The Networks and Information Integration Net-Centric Checklist

The NII Net-Centric Checklist (see Table 3.2 in Chapter 3) was prepared by the OSD to help DOD program managers understand what attributes are needed for acquisition programs that must fit into the emerging network-centric system. It was also prepared to help ensure that programs are aligned with the DOD's Net-Centric Data Strategy. This checklist is a living document that will be updated as needed. The committee reviewed its most recent version to determine whether, in the committee's view, the checklist is a help or hindrance to the Navy and Marine Corps as they build FORCEnet. Following are the committee's observations on this question.

The checklist is a fairly demanding document, containing perhaps a hundred or more questions, each of which will require some thoughtful effort to answer. Typical questions are, "Describe how the visible data assets are made available to other users outside the Community of Interest with a need for the data" and "Does the service offering depend on or use any other services provided by a different program or service provider—if so, explain how this works."[7]

The NII checklist covers a wide variety of topics, ranging from questions such as the example given above about how data will be made available to external applications, through scalability of services, to specifics of information assurance and datagram transport. Specific protocols are called out for compliance (e.g., IPv6 and XML), as are broader umbrellas such as the Joint Technical Architecture and JTRS specification. No checklist is ever complete, of course, but this document is extensive and highly detailed.

In the view of the committee, the NII checklist provides a valuable tool for the Navy and Marine Corps. It will certainly take hard work for a program manager to complete, and additional hard work will be required in the assessment of responses. However, the checklist does accurately capture many of the key attributes needed for truly open information systems. Certainly if a program manager provides deficient or questionable responses to a significant number of these questions, the Navy and Marine Corps have ample reason to doubt that the program is compatible with open systems architecture.

The committee believes that the Navy and Marine Corps should take this checklist seriously, assess programs on the basis of their checklist responses, and

---

[7]Office of the Assistant Secretary of Defense for Networks and Information Integration. 2004. *Net-Centric Checklist*, Version 2.1, Department of Defense, Washington, D.C., February 13.

favor those programs that stack up well according to this checklist over those that fare poorly. This is a valuable tool for weeding out legacy and emerging legacy systems from programs that will be most useful in FORCEnet moving forward.

### 5.3.3 Comparison of FORCEnet Compliance Checklist with Networks and Information Integration Net-Centric Checklist

The FORCEnet Compliance Checklist builds on *FORCEnet Architecture and Standards: Volume II, Technical View*. In addition to the list of standards, the Compliance Checklist includes questions on human–computer interaction (called human–systems integration), spectrum, interoperability, and documentation.

Besides the substantial overlap in the two checklists, there is also a substantial difference in approach. Whereas the NII Net-Centric Checklist focuses on capability, the FORCEnet Compliance Checklist focuses on the "What." The NII checklist is designed to understand and guide the design philosophy of the designer. The FORCEnet checklist indicates the artifacts that have to be included—for example, data types, protocols, and even aspects of documentation. Table 5.1 provides examples of the difference in approach between the checklists.

TABLE 5.1 Comparison of Two Approaches: The Networks and Information Integration (NII) Net-Centric Checklist and the FORCEnet Compliance Checklist

| Category | NII Net-Centric Checklist | FORCEnet Compliance Checklist |
|---|---|---|
| Data | Visibility, sharability, discoverability, schemas, security of, pedigree, integrity, design patterns, metadata, standards, evolution, user feedback | List of format standards, data modeling |
| Service | Open architecture, scalability of load and users, operations during outage, operations over variable bandwidth, observability of state and performance, data formats, file transfer | Time, resource locator, video conferencing |
| Information assurance/ security | Identity management, security across domains, detection and response to attacks and anomalies | Password, encryption, human security |
| Transport | Internet Protocol Version 6, multiple data (voice, video, data) on single network, quality of service, radio, fault management, detection, correction, fault isolation, diagnosis, correction, problem tracking | Transport protocols, file transfer, Internet Protocol |
| Other | | Documentation contains specific constructs, addresses certain issues |

The committee prefers the NII Net-Centric approach over that of the FORCE-net Compliance Checklist. Design philosophies consistent with an information architecture, as described below in Section 5.4, and verified by test prior to deployment have historically produced better results than have detailed checklists. The more items on the checklist, the more often it is likely to change, with resulting impacts in terms of cost and interoperability.

### 5.3.4 The Open Architecture Initiative

The committee was briefed by PEO IWS concerning its open architecture effort. The committee believes this work to be important and fundamental to FORCEnet success. The impetus for the open architecture initiative—namely, high acquisition and support costs, costly and time-consuming refresh, interoperability problems, and old architectures that are difficult to change—are similar challenges to those faced by FORCEnet. Consequently, many of the processes and policies for functional partitioning reflected in the "Open Architecture Computing Environment Design Guidance" pre-release of September 2003 are applicable to the FORCEnet architecture as well.[8] Although the scope of that document is limited to the computing environment, the critical thinking that it evidences relative to functional partitioning and interface control is exactly what is required, although missing, in the FORCEnet architecture and standards documentation.

The open architecture initiative addresses software reuse as well as refresh. As a result, the granularity of partitions to that utilized in legacy architectures increases and raises concerns—the number of boundaries to be maintained may exceed the number that can be reasonably managed, and the functional partitions may not be optimally placed. In particular, the committee believes that as long as functional partitioning supports the higher-level aggregation of interoperable functions, as suggested in Figure 5.1, it is acceptable, but the number of unique partition definitions should be minimized. The open architecture functional architecture shown in Figure 5.2, coupled with the FORCEnet information architecture described above, will, when implemented, greatly simplify the implementation of FORCEnet.

### 5.3.5 FORCEnet Executive Committee

On February 19, 2004, the ASN(RDA) issued a summary of decisions and actions from the first meeting of the FORCEnet EXCOMM. Many of the obser-

---

[8]Assistant Secretary of the Navy for Research, Development, and Acquisition. 2003. "Open Architecture Computing Environment Design Guidance," U.S. Department of Defense, Washington, D.C., September.

128

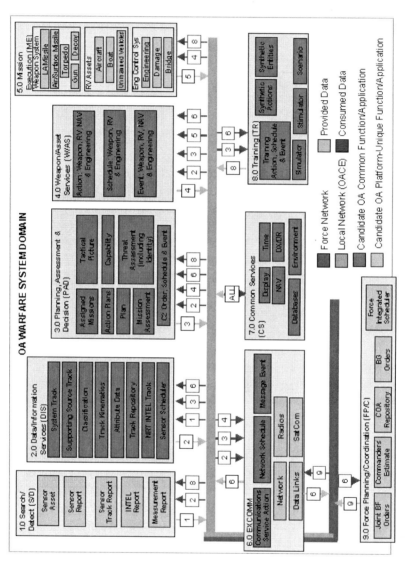

FIGURE 5.2 Open architecture functional architecture. NOTES: INTEL, intelligence; C2, command and control; NRT, near real time; RV, remotely controlled vehicle; NAV, navigation; EXCOMM, Executive Committee; SatCom, satellite communications; DX/DR, direct exchange/dead reckoning; BF, battle force; COA, common operating area; BG, battle group; OA, open architecture.

vations in the summary parallel those in this report. The decisions and actions provide a focus on the future, communicating an urgency for FORCEnet implementation that was not apparent in many of the briefings that the committee received during the period of this study. The decisions went beyond those reflected in the SPAWAR FORCEnet Business Strategy Version 1.0 and expanded the number and scope of FORCEnet pilot programs. Near-term pilots for the Marine Corps, the DCGS, and the JFN were added, and a decision was made to reallocate funds to support FORCEnet implementation.

The EXCOMM decisions, vigorously implemented, will do much to make FORCEnet a reality. Early evidence of outputs from the FORCEnet baseline process indicates that good progress is being made in identifying the tools and data necessary to implement the ASN(RDA)'s direction.

The ASN(RDA) is to be commended for the leadership displayed and actions taken. One concern is the relatively low level of participation by fleet and Marine Corps warfighter leadership in the FORCEnet EXCOMM. The actions necessary to move FORCEnet forward in a fixed-resource environment could impact near-term fleet readiness and must be accomplished in partnership with the warfighters.

## 5.4 SYSTEM ENGINEERING CONSIDERATIONS

### 5.4.1 The Big Picture

FORCEnet and the fighting units and command-and-control structure that it supports are all subsystems of a joint battle force. Systems engineering is a process for allocating functionality to subsystems that are bounded by system architecture so that the probability of mission success is optimized within available resources. A battle force performs three major functions: it manages battle, dominates battlespace, and sustains control over the battlespace over time. FORCEnet functionality is a subset of battle force functionality that can contribute to battle management, battlespace dominance, and sustainability. FORCEnet cost and contribution to battle management, battlespace dominance, and sustainability should provide a basis for implementation decisions. As a subsystem, FORCEnet must interface seamlessly with the remainder of the force while increasing the probability of mission success more than alternative investments. Understanding and defining the interfaces between what is in the FORCEnet subsystem and what is outside of it will be an ongoing process. This top-down view of FORCEnet, together with the bottom-up work that is being done at the information architecture boundaries, is necessary to explain and quantify the warfighting value. Demonstrations of value with frequent delivery of incremental capabilities are an important way to validate systems engineering practices and maintain support for progress over the long term.

### 5.4.2 What Are the Independent Variables?

The first and perhaps most important job of the system engineer is to determine the few independent variables that control the overall effectiveness of a design. The reaction time of the battle force as a system is proposed as a primary, design-driving independent variable for FORCEnet. Given a fixed set of assets, the reaction time of the battle force as a system is directly related to the validity and timeliness of the information that FORCEnet will gather and deliver to decision makers and executers. The volume of battlespace that a fixed force can dominate is determined by the reaction time of the force. Therefore, the mission effectiveness of FORCEnet is directly related to the force's reaction time and is measurable as described below in Section 5.5. The components of reaction time may change in definition for different mission areas but should always be readily measurable. The programmatic implication of the description of FORCEnet mission effectiveness given above is that the FORCEnet investment portfolio that minimizes force reaction time will be the same one that maximizes mission value.

### 5.4.3 Subnetwork Coupling

It is useful to think of the "net" part of FORCEnet as a collection of subnetworks that share the same transport layer: a collaborative sensor/intelligence subnetwork requiring high volume, a command-and-control subnetwork requiring high security, and a fire-control subnetwork requiring deterministic latency. Each of these subnetworks might themselves have subnetworks that are part of a community of interest. The fire-control subnetwork has attributes and independent verification and validation (IV&V) requirements characteristic of a manned-safety-rated system. Thus, the attributes are expensive to achieve and take a long time to test to adequate levels of confidence. Inserting a human into the interface between the command-and-control subnetworks and the fire-control loops provides a means to couple the two functions loosely and thus reduce the IV&V burden.

### 5.4.4 FORCEnet Scope and Return on Investment

However FORCEnet ends up being delivered in terms of product and capability, there will be interfaces to maintain. Once it is known what battle force functions are to be assumed by FORCEnet, the set of interfaces that must be maintained to allow the horizontal integration of display and data and the vertical integration of subnetwork functionality by mission area will be understood. It is already known that some of the interfaces that must be controlled are internal to ongoing acquisition programs and directly impact the architecture of the software and hardware that comprise these programs. In effect, some existing systems could be torn apart and their functionality rearchitected to support horizontal integration. Or, a decision could be made that the value of the information that these systems generate or the functions that they perform do not improve the probability of mission success enough to be shared across the force. Or, a decision could be made that integrating other system information or functionality with the legacy system does not improve mission success sufficiently to justify the cost. In some cases it may be possible to use software wrappers to extract the desired data and functionality from a system so that it can contribute to mission success. Although this approach adds maintenance cost since there are more lines of code to maintain and that can degrade performance, it can often be done more quickly and with less up-front investment than is required for wholesale re-architecture.

The committee suspects that a large percentage of legacy systems would not pass a cost–benefit analysis for inclusion in FORCEnet, and that fewer yet would have a high return on investment. Integration should begin with legacy systems that have a high return on investment and new developments. A rigorous and disciplined return-on-investment process by a nonadvocate will be required in order to achieve maximum return on investment and to control total ownership cost. The point here is that interfaces are difficult and expensive to maintain. The complexity and latency of evolutionary development grow exponentially with the number of unique interfaces that must be maintained, and a process must be in place to ensure that value to the fighting force can trump individual program advocacy.

### 5.4.5 Change Management

Change management for FORCEnet will require an engineering authority at a higher level of system than has ever been achieved in the DOD. Change management is the most important engineering discipline relative to maintaining information security. A construct for the way that change will be managed for FORCEnet functional partitions and standards in the context of the GIG and joint force must be developed in concert with the FORCEnet designated approval authority.

### 5.5 MANAGEMENT OF OPERATIONS

It is expected that one or more control facilities will be required to manage the performance of the FORCEnet infrastructure. Based on the methods for managing the U.S. telecommunications infrastructures as well as satellite networks (e.g., the Defense Satellite Communications System), the management centers will be highly automated, subject to human monitoring and override. Control facilities should have the capability to measure and manage to specified metrics. Automatic network measurement and management tools abound; however, tools for measuring mission effectiveness in a network-centric force are not yet mature. The flexibility of a network-centric force can provide unprecedented options to the operational commander if the tools to monitor actions and outcomes are in place. A minimum set of operational metrics for FORCEnet is provided below.

### 5.5.1 Recommended Metrics for Communications Transport

Following are the metrics recommended for communications transport:

• *Availability*—High (multiple nines) for time-critical warfighting functions versus lower for routine messages. Availability of communications should be viewed from the end user's perspective and managed to mission priority.

• *Packet round-trip time*—Packet round-trip time is a measure of the congestion of the network and is an end users' view of quality. It is impacted by bandwidth, routing efficiency, and firewall settings. It must be tightly managed for time-critical messages. Note that for stream data such as ISR imagery, it may be necessary to maintain circuit-switched rather than packet-switched subnetworks.

• *Packet (or message) loss percentage*—Packet (or message) loss percentage is an indication of problems that will ultimately impact packet round-trip time and customer satisfaction as available bandwidth is consumed. Usually every packet or message lost has to be resent. Packet loss over 1 percent is a problem that can often be corrected with rerouting. Some critical functions will require acknowledgment such as pulling posted information or acknowledging a command for which a commercial IP such as Transmission Control Protocol is appropriate. Other data sharing, as for collaborative sensor networks, may not require that every data message be received—User Datagram Protocol, for example, may be the appropriate IP in this case.

### 5.5.2 Recommended Metrics for Warfighting Effectiveness

The following three interoperability metrics for the antiair-warfare area are an example of those recommended for measurement of the warfighting effectiveness of FORCEnet. The sum of the three metrics is a representation of battle force reaction time. The definition of the elements that make up reaction time will vary from warfare area to warfare area, but in each case their sum should be a representation of battle force reaction time and should include the latency of critical mission-information sharing, the latency generated by command-and-decision-tool interoperability issues, decision-maker latency to analyze and act on information presented, and the length of time it takes to communicate the execution decision to the executor. The sum of these parts should decrease as battle force interoperability improves with FORCEnet. Faster is always better. As VADM Arthur Cebrowski, USN (retired), once said, "Show me a person who does not

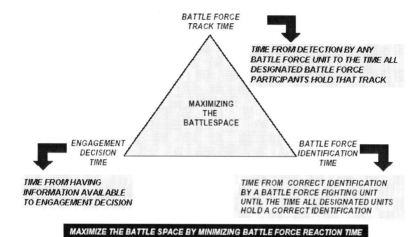

FIGURE 5.3 Battle force reaction time. SOURCE: Joseph Cipriano and Jerry Krill. 1999. Battle Force Interoperability and Net-Centricity, *Defense Systems and Equipment International*, Applied Physics Laboratory, Johns Hopkins University, Laurel, Md., pp. 261–270.

think speed is important, and I will show you someone who has never been shot at."[9] The interoperability metrics for the antiair-warfare area are these:

• *Battle force track time*—The time from the first detection of an unknown by any battle force unit to the time when all designated battle force participants hold the detection and track, if applicable (connectivity availability and the effectiveness of correlation algorithms are measured).

• *Battle force identification time*—The time from the first correct identification of a detection by a battle force unit until all requiring units hold a correct identification (interoperability of identification algorithms and message round-trip time are measured).

• *Engagement decision time*—The time from when sufficient information is available to the time when an engagement decision is made (quality and completeness of data presentation and operator training are measured).

When battle force reaction time is degraded, more assets must be allocated to defend the battlespace (see Figure 5.3). If a FORCEnet capability does not improve battle force reaction time, it should not be deployed.

---

[9]William D. Eggers. 2005. "On Point," *Government Technology's Public CIO Magazine*, February.

## 5.6 FACILITIES

In classical systems engineering, the subsystems are typically brought to-gether in a laboratory setting with scripted drivers to ensure proper integration via strict test processes. This integration buildup, especially for subsystems delivered by different, physically separated organizations, must be carefully planned and well resourced (e.g., Aegis integration and testing at Moorestown, New Jersey; Aegis Computer Center at Dahlgren, Virginia; and Aegis Combat System Center (ACSC), Wallops Island, Maryland. Early in the integration and testing of both Navy Link 16 and Cooperative Engagement Capability, a limited testbed link between ACSC/Wallops and Fleet Combat Direction Systems Support Activity at Dam Neck, Virginia, allowed these networks to operate with Aegis and carrier Advanced Combat Direction System combat systems with scripted and live sce-narios. These proved critical to the early success in formal testing of these systems.

However, there was no mechanism for more extensive testing other than onboard ships of a battle group. Since each deploying battle group configuration is different, this resulted in key interoperability issues that were not discovered or addressed until a battle group began predeployment workups. Instrumentation for problem detection and time for correction prior to deployment were limited. Increasing numbers of such events in the late 1990s resulted in the establishment by the Naval Sea Systems Command (NAVSEA) of a DEP and a battle-group-specific deployment-minus-30-month countdown configuration management and test process for combat and C4I systems planned for deployment.

Thirty months prior to deployment, the fleet commander identified the plat-form assets that would make up a battle group. A battle group action officer was then assigned by NAVSEA to manage the configuration and certification testing of the combat and C4I systems on the designated platforms in the DEP using scripted scenarios in concert to drive real-time interactive testing. The battle group staff observed the DEP testing and developed concepts of operations and communications plans based on the observed capabilities and limitations. After critical deficiencies were corrected, the battle force system engineer would cer-tify a particular battle group configuration as safe and effective for deployment. The DEP proved able to baseline the systems of a deploying battle group and to ensure correction of key deficiencies. However, the DEP does not have sufficient capacity to support the originally intended additional uses, such as the early testing of concept models and prototypes and the verification of engineering compliance and interoperability during the development cycle for systems intended for eventual deployment. It has been consumed assuring near-term readiness.

The DOD has identified the need for a JDEP to leverage the demonstrated value of the Navy's DEP in preparation for joint operational deployments, as well as for the verification and risk-reduction activities earlier in the development cycles of joint and Service-specific systems in a joint setting. However, the definition and focus of the JDEP have been slow in realization of their intended capabilities. Further, the JDEP is not currently anticipated to be sufficient in scale and availability to serve anticipated Navy needs for FORCEnet according to the process identified above. Given the experience with the DEP, the Navy should play a lead role in realizing an extended joint DEP by first extending the Navy DEP.

To reiterate, the extended Navy DEP for FORCEnet would provide for the following:

- Early exploration of concepts and concepts of operations against future threat scenarios,
- Integration testing and risk-reduction evaluations of prototypes and models of conceptual FORCEnet elements,
- Baselining of configurations prior to critical field experiments and data collections,
- Verification of compliance with FORCEnet standards and architecture boundaries and interfaces,
- Integration and test buildup to operational tests,
- Baseline testing and fixes prior to deployments,
- Training to new capabilities and doctrine, and
- Problem replication and fix verification for deployed systems.

The extended DEP would provide the following additional features relative to the existing DEP:

- Both simulated and actual networks and routers/hubs,
- Wrap-around network-wide simulation drivers of the networks as well as system applications,
- Interfaces to the eventual JDEP,
- More nodes especially for the contractors for verification during development as a risk reduction activity,
- Connection to facilities such as Aegis ASCS/Wallops for live radar and weapon systems,
- Interface to corresponding national assets test sites, and
- Potential to connect to predeployed combatants for further verification testing.

## 5.7 VULNERABILITIES AND SOLUTIONS

Every time new assets are added to a networked system, new vulnerabilities can be added as well. To offset this risk, networks offer increased opportunities for implementing security in depth. The FORCEnet architecture must support security at access points, and it must do so in the transport and application layers as well as at the data layer. Concepts and mechanisms for maintaining access controls to network, applications, and data must be developed and built in up-front.

Whenever needed by warfighters—and especially at intervals of high war-fighting stress—the network and its information must be available and trustworthy. Operators must be able to rely on the network and its information without constant concern that the information they are using has been poisoned by deliberate acts of an adversary, that an adversary is inside the network, or that the entire network might suddenly collapse spontaneously or under enemy attack.

Although there are many possible ways to categorize the vulnerabilities of large, networked systems, the major dangers fall under the following headings:

- Unexpected fragility,
- Worms and viruses,
- Insider threats,
- Threat of poisoned data, and
- Denial-of-service attacks.

As sets of plausible attacks gradually become apparent, information assurance mechanisms and policies can be developed to counter them. At present, however, there is a noticeable lag between new attacks and the implementation of corresponding defenses, since fresh software is constantly being installed and its vulnerabilities only become apparent over time. A key challenge will be to reduce this lag until it is small enough so that adversaries find it infeasible to mount attacks during the undefended interval.

The committee notes that information assurance policies and technologies provide protection against risk at some cost (burden), in terms of both efficiency and restrictions on operator activities. Do the policies need to be applied the same way in all circumstances? How may a commander control the risk or burden trade-off to suit changing situations? All of these are currently open issues needing attention.

### 5.7.1 Unexpected Fragility of Complex Systems

As strange as it may seem, possibly the greatest danger to FORCEnet may be its own spontaneous collapse or malfunction, without any obvious external cause. In the past, large-scale distributed systems, such as networks and electrical systems, have often proved to be surprisingly fragile. Seemingly trivial accidents in a minor part of the system can quickly cascade into overall systemic failure. This inherent fragility should be of great concern to the Navy as it continues to rely more and more on large, interconnected information systems.

Several well-known cautionary examples are listed below. It is important to recognize that in each of these cases, a stable, well-engineered, well-run system underwent a rapid and totally unexpected collapse. Although triggered by trivial, transient events, total system outages often lasted for days.

- *Northeast power outage of 2003*—Two discrete and unrelated minor events led to the loss of power to 50 million customers for 3 days.
- *AT&T nationwide failures, 1990 and 1998*—Software flaws introduced in system upgrades resulted in nationwide loss of AT&T long distance for 9 hours in one case and took thousands of businesses offline for 6 to 26 hours in another.
- *Baltimore tunnel fire in 2001*—A chemical spill and fire severed three major fiber-optic lines for the East Coast. Telephone call centers from Maryland to New Jersey lost service for half a day.

Although the Navy and Marine Corps do not currently appear to be particularly vulnerable to such catastrophes, it does not take too much imagination to picture how such cascading failures might affect the Navy once all naval platforms are networked. What if router software for all platforms in a battle group were upgraded to the same release, and then a minor anomaly led to a 24-hour outage of communications across the battle group? It happened to AT&T twice. What if a single fiber cut in the continental United States led to day-long outages and slowdowns for accessing Web databases or for important centralized services such as chat rooms? It happened to the largest, best-run Internet providers in the United States. What if simple viruses and worms entered the Secure Internet Protocol Router Network and brought e-mail to a crawl—or even clogged essential communications links to tactical platforms? It happened to NMCI.

Nonspecialists may think that these are remote and unlikely situations, but experience has shown precisely the opposite. Large and complex systems are surprisingly fragile. They require great care in their design and operation.

### 5.7.2 Coping with Vulnerabilities

The previous subsection listed a number of technical dangers that will threaten FORCEnet for the foreseeable future. These dangers cannot be eliminated, but imposing the following four requirements on FORCEnet and its operations can ameliorate them:

- The network infrastructure must be extremely robust, with sufficient redundancy and diversity to adapt to losses of component portions. There must be specific plans and processes in place for adapting to each potential loss.
- Since a loss of a portion of the network is very likely to reduce the capacity of the network and of the communications capability to some of its

users, each adaptation plan must include the changes in operations that are required to adjust to the reduction in communications, and training must include the operations at each level of reduced capability.

• Since there is a significant likelihood that some ends of the network at some time will lose connectivity totally, sufficient local capability must be provided to allow effective operation in such a mode for a period of time.

• A "one-size-fits-all" solution of hardware or software is not recommended, since one vulnerability in a common component can lead to the complete crippling of the entire system. Thus, some level of heterogeneity is needed to help combat worms and viruses.

## 5.8  FINDINGS AND RECOMMENDATIONS

### 5.8.1  Findings

Following are the findings and observations of the committee with respect to FORCEnet architecture and system design:

• The process and tools for translating FORCEnet operational concepts into products, services, and warfighting capabilities have yet to be fully developed. Systems engineering is a process for allocating functionality to subsystems that are bounded by a system architecture. The current SPAWAR architecture document is difficult to follow, overspecifies standards, and provides incomplete identification and specification of architectural boundaries to support FORCEnet systems engineering. NAVSEA open architecture work reflects a process for establishing boundaries and partitioning functionality that is representative of what is required for FORCEnet. The system engineering expertise, disciplines, and lessons learned that NAVSEA has developed in support of open architecture would help fill the gaps in the current FORCEnet systems engineering process.

• The complexity of the complex system providing FORCEnet capabilities makes executing systems engineering a great challenge. The number of unique interfaces that must be maintained need to be carefully selected and kept to an absolute minimum, or evolution will be hindered by expensive and lengthy integration and testing. One way to do this is to require that systems must partition common functions in a common way. Web service architectures are a good example of this principle.

• The warfighting value of FORCEnet capabilities and speed of implementation are paced by bandwidth availability, and yet little has been done to estimate bandwidth requirements and to develop solutions to support the implementation of a network-dependent force.

• There has been little attempt to characterize how FORCEnet will function in terms of network management, data flow, traffic control, nodal performance, or data access. This information is required to engineer the FORCEnet network-management system.

• It is not evident that existing facilities are adequate to support the engineering, training, and phased deployment of FORCEnet capabilities. With the fast pace of change in C4I and IWS, it remains necessary to perform high-fidelity land-based testing to ensure interoperability prior to deployment as is currently done in the NAVSEA DEP. Current facilities are not capable of handling an increased workload.

• The NII Net-Centric Checklist provides an excellent tool for the early evaluation of systems proposed for inclusion in FORCEnet and the GIG.

• The FORCEnet EXCOMM is a laudable start at a governance process for FORCEnet, but it does not have the proper level of representation from the fleet to function effectively with the challenges ahead. The leadership of ASN(RDA) in this effort is vital and commendable.

• The FORCEnet network controls do not provide the capability to meter and to prioritize data flow across boundaries, and FORCEnet behavior models are not developed to project performance and to support sensitivity analysis.

• The FORCEnet architecture does not provide redundant and diverse communication paths to guard against network vulnerabilities and degradation—for example, by furnishing alternatives to satellite communications for the last mile.

• The reaction time of the joint force to sensor input is not a design-driving requirement for FORCEnet or an evaluation factor for the prioritization of activities.

• A FORCEnet roadmap has not been developed, showing the schedule for design and analysis deliverables and the incremental delivery of capability.

### 5.8.2 Recommendations

Based on the findings presented above and on the issues described in this chapter, the committee recommends the following:

• **Recommendation** for the CNO and the ASN(RDA): Take measures to strengthen the FORCEnet architecture in terms of its ability to represent overall structural relationships among force components. To this end, the CNO and ASN(RDA) should designate NAVSEA, drawing on its open architecture experience, as having a major role in developing the FORCEnet architecture, particularly as pertains to its representation of invariant boundaries and the ability to allocate functionality. Furthermore, SPAWAR and NAVSEA should be directed to specify the technical interrelationship between the FORCEnet architecture and the combat systems open architecture.

• **Recommendation** for OPNAV: Adopt the Net-Centric Checklist of the ASD(NII) in place of the OPNAV FORCEnet Compliance Checklist, adapting it

140

if necessary to accommodate specific aspects of naval warfare. This design guidance, together with a focus on architectural boundaries, should help promote the development of FORCEnet architectural products.

• **Recommendation** for the ASN(RDA), with the support of the systems commands and the relevant PEOs (primarily PEO C4I & Space and PEO IWS): Develop the capability necessary to effect FORCEnet systems engineering. Very high standards, commensurate with the challenge, should be set for the systems engineering staff, who can come from the systems commands, program offices, and outside sources, as necessary. This systems engineering capability would work directly in support of any organization developed to integrate the Navy's program formulation and acquisition functions more closely (as discussed in Section 4.8.5).

• **Recommendation** for the ASN(RDA), the systems commands, and the relevant PEOs (primarily PEO C4I & Space and PEO IWS): Pay particular attention to the following in establishing the FORCEnet systems engineering capability:

—Instituting a change management authority responsible for the full set of FORCEnet functional partitions and standards. The decisions of this authority will affect a broad range of naval programs, unprecedented in any prior DOD systems engineering. This authority is key to maintaining the integrity of overall FORCEnet capabilities.

—Providing for the frequent delivery of system capability (e.g., in 6-month increments). This will reinforce the value of the systems engineering process and is in keeping with the need for evolutionary development (discussed, e.g., in Section 7.3.2.3).

—Achieving a mission focus in the analysis and allocation of functionality. For example, each mission can be characterized by a few key variables (e.g., battle force tracking and identification times in antiair warfare) that should be optimized in system design.

—Establishing a rigorous process, independent of individual programs, for recommending the future course of legacy programs—phaseout, retention as is, upgrading, or merger into another program—based on the mission utility of each program.

—Establishing means, involving both process and technology development, to recognize and deal with the vulnerabilities and fragilities that could cause significantly degraded overall capabilities.

• **Recommendation** for the CNO and the CMC: Establish a joint operations research capability for complex distributed systems. The operations research organization would be resourced to develop the concepts of operation, design reference missions, and performance models necessary to validate and prioritize operational requirements, including bandwidth requirements, for a network-centric force.

• **Recommendation** for the ASN(RDA), in conjunction with the systems commands and the relevant PEOs (primarily the PEO C4I & Space and PEO IWS): Develop a FORCEnet DEP by generalizing the concepts and approaches used in the current DEP. Since the ongoing joint distributed engineering plant effort does not appear adequate to meet FORCEnet needs (e.g., in terms of scale), the Navy should play a lead role in realizing an extended JDEP by first extending the Navy DEP.

• **Recommendation** for the ASN(RDA): Invite the fleet community to provide a senior flag officer to participate in the FORCEnet EXCOMM decision process.

# 6

# Science and Technology to Support the FORCEnet Information Infrastructure

## 6.1 OVERVIEW AND BACKGROUND

Because FORCEnet has no fixed end state but is subject to continual innovation, it is not possible to establish a fixed set of science and technology (S&T) investments required to enable successful implementation of the FnII. Figure 4.2 in Chapter 4 of this report depicts the view of the CFFC,[1] of this innovation process, showing Sea Trial as the center of a process in which warfare challenges lead to new concepts. These concepts drive both materiel and nonmateriel innovations that are tested in Sea Trial, iteratively refined, and, when implemented, lead to new operational capabilities.

In Figure 6.1, the committee places this construct in a larger context. (The processes added by the committee are identified by the shaded boxes and the dotted lines in Figure 6.1.) The chart closes a loop to account for the Network-Centric Operations Capability Vision, including evolving threats and new warfare challenges being affected by capability gaps and indicates that technology gap analysis can motivate science and technology programs that will lead to technology that can help close the gaps in operational capabilities. A path for spiral experimentation has also been added. This path connects the "New Operational Capability" box in Figure 6.1 with the "NCO Capability Vision and Evolving Threats" box. In this way, the new operational capabilities can use spiral experimentation to determine how well they perform with respect to the evolving

---

[1]ADM Robert J. Natter, USN. 2003. "Sea Power 21 Series, Part VIII: Sea Trial: Enabler for a Transformed Fleet," *U.S. Naval Institute Proceedings*, November, p. 62.

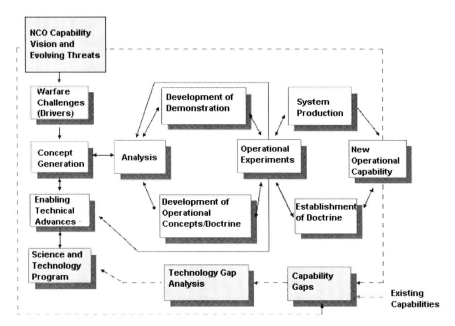

FIGURE 6.1 Recommended augmented process for identifying technology gaps in operational capabilities. Adapted from ADM Robert J. Natter, USN, 2003, "Sea Power 21 Series, Part VIII: Sea Trial: Enabler for a Transformed Fleet," *U.S. Naval Institute Proceedings*, November, p. 62.

threats indicated by the Director of Naval Intelligence, the intelligence community, and the combatant commanders.

The committee had access to the following material: gap analyses of operational capabilities performed by N704, a translation of these gaps to S&T needs performed by N706,[2] and parallel but not entirely consistent requirements generations and S&T shortfall lists produced by NETWARCOM and its OAG. The committee also had access to work by ONR identifying critical enabling technologies and to SPAWAR's Technology Framework for FORCEnet.[3] The committee also studied aspirations of the ASD(NII) for the GIG, and made its own assessment of the technical challenges facing naval implementation of FnII while leveraging GIG capabilities (discussed in Chapter 3, Section 3.6) and complying with GIG requirements.

---

[2]N704 is the FORCEnet Integration and Assessments Branch for N6/N7; N706, the former Science and Technology Branch, no longer exists.

[3]Space and Naval Warfare Systems Command. 2003. FORCEnet Government Reference Architecture, Version 1.0, April.

To deal with evolving statements of desired operational capabilities, and recognizing the evolutionary process of FORCEnet implementation, the committee abandoned the attempt to set hard performance goals for FnII technology. Instead, it integrated the ONR taxonomy[4] and its own analysis of GIG challenges into a list of eight FnII critical technologies (see below). Subsequent sections in this chapter describe the challenges in each, but the establishment of metrics that must be met at a particular date need to await agreement on the operational capabilities desired for that date and for better modeling and simulation tools to accurately assess the effect of FORCEnet performance on the warfare effectiveness of the Navy.

The committee examined several documents, as mentioned, and from those documents it identified eight critical FnII functional capabilities, listed below. These capabilities, used as the organizing basis for discussion in this chapter to highlight potential capability gaps and associated S&T shortfalls, are as follows:

- Reliable wideband mobile communications;
- Information management (including COP);
- Situational awareness and understanding;
- Information assurance;
- Modeling and simulation;
- Dynamic composability and collaboration;
- Support of disadvantaged user-personnel, platform, or sensor; and
- Persistent intelligence, surveillance, and reconnaissance.

## 6.2 ENABLING TECHNOLOGIES AND FUNCTIONAL CAPABILITIES

### 6.2.1 Reliable Wideband Mobile Communications

#### 6.2.1.1 Communications Overview

In today's naval forces, communications with moving platforms and personnel are characterized by intermittent connectivity and low data rates. Today, most ships have only satellite capability and achieve data rates of less than 100 kb/s, whereas larger ships, such as carriers, can achieve data rates of multimegabits. Satellite connectivity for small ships is often in the area of 80 percent. The primary limiting factor appears to be antenna blockage, but other possibilities could be electromagnetic interference (EMI) and tracking problems during dynamic maneuvers. Submarines, which must put an antenna on or above the sur-

---

[4]Office of Naval Research. 2002. "Taxonomy of Technology Limitations to Support the Five Enabling Functions Required for Navy Network Centric Operations," Arlington, Va. Available at http://www.onr.navy.mil/02/baa/expired/2003/03_007/default.asp. Accessed July 24, 2004.

face, and dismounted troops have more serious data rate and connectivity problems. Operations in the future network-centric environment of FORCEnet (see Figure 6.2) will require much higher data rates (fleet representatives have estimated requirements to be as high as 50 Mb/s per large-deck ship) and more-robust connectivity. Simultaneous connectivity to satellites, sensor nodes, airborne relays, and other ships will drive antenna requirements. Furthermore, maintaining an Internet type (e.g., IP-based) network while nodes and users are moving is a significant technological challenge.

### 6.2.1.2 Communications Technology Challenges

The following subsections discuss key communications technology challenges that must be addressed if FORCEnet is to achieve the vision of full network-centric operations.

***Communications Links and Apertures.*** The difficult shipboard environment for radio-frequency (RF) apertures makes it a high-priority area for future improvement. An example of the antenna layout on a typical ship today is shown in Figure 6.3. The figure illustrates the many trade-offs that need to be considered in planning communications antennas on a ship. The larger the antenna, the greater its throughput, but larger antennas also have larger cross-sections and so their detectability is larger. Also, pointing accuracy and the amount of topside space required increases with the growth in antenna size.

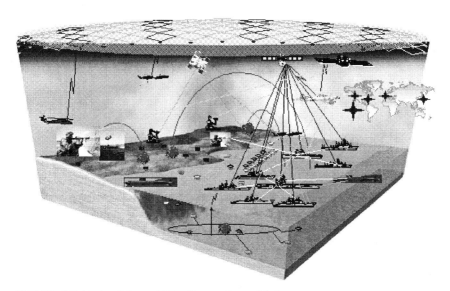

FIGURE 6.2 Notional future FORCEnet nodes and links.

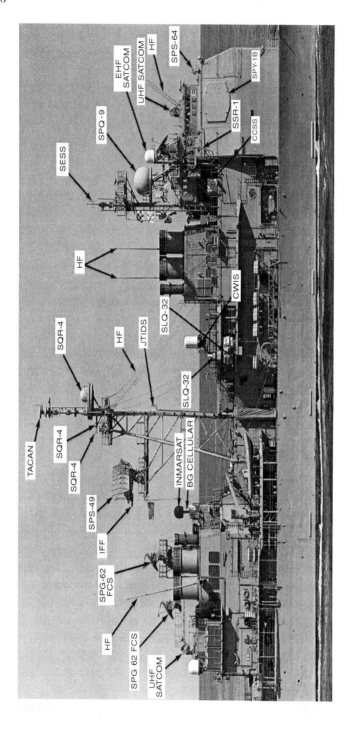

FIGURE 6.3 Example of the antenna layout on a typical ship today: antennas resident on a U.S. Navy cruiser, circa 1996. NOTE: A list of acronyms is provided in Appendix C. SOURCE: CDR J.J. Shaw, USN, Head, Space Section (N611), "Transformational Communications Architecture Overview" presentation, August 11, 2003, briefing.

Typical improvements addressed are (1) the sharing of apertures among different functions (radar, communications, and so on) and frequencies and (2) multiple simultaneous (or agile) beams to allow communications with multiple, independent nodes. Operating at higher frequencies (Ka band and above) provides improved performance for a given transmitting and receiving aperture size. This, however, increases the pointing and tracking challenge owing to the reduced beam widths. Optical frequencies should be investigated for communications from UAV relays or satellites to ships in order to provide the increased bandwidth when allowed by the atmospheric environment. While deformable mirrors are potential technology solutions for ameliorating the distortion caused by the atmosphere, serious issues related to atmospheric scattering in the marine environment remain, without clear means of being overcome.

As the Navy moves toward the distributed nature of a network-centric FORCEnet capability, ships will need to be able to track multiple signal nodes simultaneously. An example, taken from the LCS concepts of operations, is shown in Figure 6.4. The need for multiple agile beams indicates a phased array

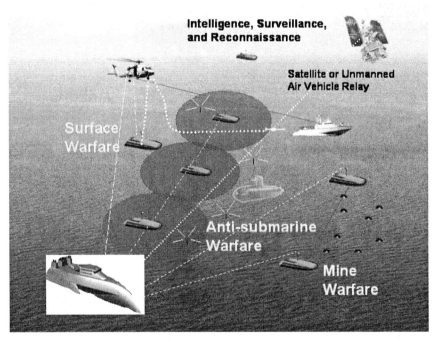

FIGURE 6.4 Concepts of operations overview for the Littoral Combat Ship and its distributed off-board systems. SOURCE: Navy Warfare Development Command. 2003. *Littoral Combat Ship, Concept of Operations Development*, Newport, R.I., February.

as one possible solution. Several programs supported by ONR are addressing these antenna issues.

Automated, intelligent management of link characteristics, including beam pointing and tracking, jamming, spectrum usage, and atmospheric environmental effects such as rain attenuation and fading, would be needed to maintain the robustness of link performance. Note that Figure 6.4 points out the need to consider alternatives to satellite relays, such as unmanned air vehicle relays. These alternatives are required because of possible shortages of satellite capacity, owing either to congestion or to adversary action, and to blockages of the line of sight from a shipborne antenna to a satellite.

***Network Quality of Service and Resource Management in a Military Context.***
In today's commercial Internet, there is little or no capability for allocating QoS among various classes of traffic. The result is that all messages have the same priority, and when there is a heavy demand on the network, all users experience the same degradation. In a military context, high-priority traffic needs to take precedence over low-priority traffic; otherwise a heavy demand of low-priority traffic could preempt the higher-priority information. The transition to IPv6 from IPv4 will provide a tool, but more work will be needed in order to enable reconfiguration of the network infrastructure in response to varying military missions. Monitoring and control of the infrastructure from the network level and down to the link level will be required for enabling response to the time-varying needs. The monitoring and control should be automated to the extent possible, especially when latency is a critical requirement. Although several programs address monitoring and control, the committee finds no comprehensive systems-level effort directed at the total problem.

***Automated Networking in a Dynamic, Mobile Environment.*** The ability of moving personnel and platforms to continuously maintain the data rate and connectivity necessary for the achievement of their assigned missions is a significant technological challenge. The standard protocols used in the commercial Internet work well in the fixed infrastructure, but when the users and, in particular, the nodes and hosts are moving, special protocols are required.

If only the hosts (e.g., users with laptop computers) are moving and the routers are static, the situation is easily handled by the Mobile IP without placing significant burden or design changes on existing routing protocols, such as Open Shortest Path First. A more challenging situation arises when the routers are also moving, as would be expected in a dynamic battlefield situation involving ships, troops, and UAVs. Without new routing protocol designs, the system would lose track of the user locations (i.e., which router they are using), connections would time-out, and connectivity would be lost. This type of network is referred to as a Mobile Ad-Hoc Network (MANET).

The MANET Working Group of the Internet Engineering Task Force has been working on developing standard routing protocols for MANETs. Today, there are several MANET protocols operating at the IP layer—for example, Ad hoc On-demand Distance Vector (AODV) and Optimized Link State Routing (OLSR)—but they are still classified as experimental. The working group has not come up with a single solution, because it was thought that too many unknowns exist and the solution may be situation-dependent.

Some protocol designs keep constant track of the locations of all users, so that if one wishes to communicate with another, the path is already known. Such solutions, known as proactive (such as OLSR), have high overhead but low latency. Other designs, known as reactive (such as AODV), determine a path only when needed, resulting in lower overhead but higher latency. Also, there is some uncertainty on the scalability of these solutions to hundreds of nodes or more. Although some implementations of MANETs exist and some experiments have been done (e.g., the Army's FCS program), insufficient information is available today to allow systems engineering trade-offs.

Another challenging problem area arises when some or all of the links of the network are not constantly connected, but suffer dropouts for varying periods of time. With today's protocol designs, such behavior causes large numbers of repeated transmissions of packets and possible crashing of the network. The Delay-Tolerant Networking (DTN) Research Group (of the Internet Research Task Force) has been addressing this area for some time, but, as is the case with MANET, solutions are immature. The Defense Advanced Research Projects Agency (DARPA) has initiated a new program, Disruption-Tolerant Networking, to address this area.

Overlaying the issues of routing protocols for mobile and disruptive networks is the issue of resistance to adversarial attack—that is, information assurance specific to the network. A robust network must not be susceptible to data corruption, corruption of routing, or saturation of the network with garbage traffic (denial of service). It must also be resistant to traffic analysis (i.e., to revealing who is sending information to whom). Of course the way to protect such information is to encrypt it, but if the headers containing the protocols are encrypted, they must be decrypted at each router if the routers are to take action based on them. Thus, there appears to be a trade-off between security and performance in a dynamic network, especially within the concept of a "black core," that is, the part of the network with the highest security.

The future FORCEnet will have the characteristics of both MANETs and DTN; emphasis should be placed on continued research in both of these areas. On top of this will be the need for a level of network information assurance. A multidimensional trade-off will be needed between potentially conflicting requirements, with solutions being dependent on the specific missions, architectures, and systems designs. Today there is not enough knowledge of these areas and their interrelationships to be able to do these trade-offs. More modeling and simulation and experiments are needed to explore the solution space. Solutions for one type of mission may be different from those for other types, perhaps indicating the need for different solutions for different enclaves of users. Perhaps focusing on the LCS program would be a useful place to start.

### 6.2.1.3 Communications Science and Technology Perspectives

A considerable number of efforts are ongoing under ONR sponsorship in the area of RF antenna technology. Some work is ongoing in the optical regime, but not enough to really assess its adequacy for the future FORCEnet environment.

The committee is not aware of any efforts to support a comprehensive design of an automated monitoring and control system for FORCEnet links and networks.

The ONR, DARPA, and the Army have a number of efforts supporting MANET; DARPA is initiating an effort on DTN. As yet, these efforts are not sufficiently mature to provide performance results for dynamic networks with specific security requirements under specific missions and scenarios. The National Security Agency is developing a new security protocol, HAIPE, which will help solve some of the security issues previously discussed.

***Findings.*** Currently available technology is not sufficient to support the robust communications infrastructure needed for the long-term FORCEnet network-centric operations vision. In particular, the current technology gaps include the following:

* The capability in link and antenna technologies to provide increased data rates and beam agility;
* Insufficient quality of service and network monitoring, control, and reconfiguration to provide the necessary availability and latency for priority traffic;
* Necessary protocols in standard use to support the mobility, ability to overcome disruption, and information assurance robustness that will be needed in the future FORCEnet;
* Reliable communications technologies to reach underwater vehicles at speed and depth;
* Shared, robust, reliable, multibeam apertures, satellite relay alternatives to support communications on the move, and adaptive networks;
* Reliable high-speed communications, including optical, in the marine layer; and
* Improved antenna aperture technology for use by disadvantaged users: personnel, platforms, and sensors.

***Recommendation.*** Based upon the findings presented above and on the issues described in this section, the committee recommends the following:

• **Recommendation** for ONR: Monitor technology availability and, as appropriate, invest to sustain investigations that:

—Examine the applicability of optical frequencies for high-data-rate communications from satellite or airborne platforms to surface ships. Although the future Transformational Communication System holds promise for achieving as much as 100 Mb/s to ships at Ka band, research into optical communications could provide a hedge against a need for higher data rates in the future.

—Examine providing automated monitoring and control for FORCEnet links and networks.

—Explore the solution space for network approaches for FORCEnet mobility, disruption, and security using modeling and simulation and experimental approaches; it should particularly consider applications, such as the Littoral Combat Ship, as points of departure for this effort.

### 6.2.2 Information Management

#### 6.2.2.1 Information Management Overview

Information management encompasses a spectrum of issues critical to the implementation and effectiveness of FORCEnet. The process of information management includes the generation and manipulation of data or information in support of decision makers. By implication, the process may take different forms, depending on the decision maker's role or responsibility (e.g., command and control, strike, logistics). The information management process includes all activities related to the collection, accessing, processing, dissemination, and presentation of data or information. The process includes technical means as well as policy, procedural, and doctrinal aspects, with a focus on producing the right information (the content and quality that are needed) at the right time to satisfy mission demands. As implied, information management processes must be adjusted to satisfy specific mission drivers. A high-level summary of the contributing technologies follows:

• *Sensor management*—enterprise-mediated collection planning to maximize information value from observations of multiple areas and locations of interest by sensors most likely to satisfy mission needs, with appropriate mode, geometry, and timing; adjudication of competing demands for sensing coverage in support of all users in accord with command priorities.

• *Sensor processing and data fusion*—single and multiple sensor and source fusion to minimize uncertainty; includes temporal alignment, geospatial registration, location, tracking, identification of objects and events, and aggregation into appropriate representation of battlespace objects and events. Such processing is often referred to as Level 1 data fusion processing. In centralized architectures,

Level 1 processes are often applied to sensors and message traffic, reflecting wide area coverage, to form a common operational picture.

• *Information services*—models, pedigrees, metrics, database services, and so on to support efficient, dynamic management of information.

• *Data strategy and information dissemination*—consistent mapping of (1) data meaning and significance (classes, properties, relationships) and (2) information content, structure, and latency, to network limitations and technical capability of users.

• *User-defined visualization*—representation of information in forms appropriate to user roles; decision support to aid human cognition.

In stand-alone (platform-centric) systems, these process issues are assessed and resolved at design time. The technical aspects (e.g., data parameters, processing constraints, and information products) are embedded in the system design and tend to be modified only infrequently over the life of a platform. Typically when new requirements are imposed, needed design updates (or reengineering) are dictated by "interoperability" limitations and prove to be both time-consuming and expensive. Among cooperative platforms, protocols and procedures can be established to assure that information is exchanged within a predetermined structure and used in ways that are appropriate for particular mission goals.

**6.2.2.2 Information Management Technology Challenges**

In network-centric operations, the information management process must work across all node components of the network in a fashion that is seamless and adaptive to command direction. This suggests that all nodes must make all their contributing elements of the information management process transparent to network command and control. Further, network environments will be characterized by high volumes of data or information supporting a diverse set of users and mission goals. In such an environment, there is evident need for underlying consistency in the description of information and for automation in the application of tools to enable dynamic and efficient information management. Without such automated tools, the network will become bottlenecked by delays and capacity limitations caused by humans engaged in futile efforts to resolve information conflicts and inconsistencies. This is an issue of information integrity—the requirement that information, as it propagates around the network, be processed and interpreted in ways that are mathematically and logically consistent with source sensing characteristics and intermediate processing updates.

Common problems occurring in today's battle management environments, which will be dramatically compounded in network-centric environments, are these:

• *Trust*—lack of metadata about information source, intermediate processing, and quality to inform users;

• *Contamination*—data corruption due to redundant paths or improper processing (for example, multiple reports of the same target at different locations, or duplicative reports of ambiguous targets causing confusion about actual numbers); and

• *Utility*—lack of appropriate tools for users to process or exploit received data and information.

GIG-espoused paradigms of TPPU and OHIO carry an implication that a common discipline will be developed and invoked across all network nodes and for all users in order to ensure proper use and interpretation of information. This suggests an "information services" layer to complement the enterprise services defined under the GIG-ES and being developed by DISA in the NCES program. Possible elements of such an information services layer are common enablers, such as these:

• Metrics to qualify the information on some normative scale for consistent and (mathematically) proper usage,

• Pedigrees to identify the source and intermediate processing action (and time) to avoid redundant usage and to support validation and error correction, and

• Models to capture and share knowledge of phenomenology, platforms, sensors, and processes to guide algorithm usage and human interpretation.

### 6.2.2.3 Information Management Science and Technology Perspectives

The information management issues outlined above deal with the content and quality of the information that flows over the network. The contributing information management technologies identified above are available in varying degrees of maturity, but development is required to make those technologies suitable for network-centric purposes. Similarly, issues associated with defining and implementing information services are receiving scant attention. GIG effort being expended on enterprise core (communications and enterprise services) is not addressing these issues. To achieve the network-centric vision, FORCEnet, in conjunction with other Service activities, will need to develop technology solutions for these information management components and automation, within the communities of interest, in ways that appropriately leverage core enterprise services. Programs that provide an appropriate focus for technology development are being formulated, as in the ONR's FORCEnet S&T plan for POM-06. Those efforts are geared to COP formation and time-sensitive decision making, with early products available in FY 2008. Greater investment in this area will be needed to realize the information potential of network-centric operations.

***Findings.*** Insufficient information management technology exists for the reliable support of naval warfighting capability—including limited understanding of the information management issues (accessing, processing, dissemination, presentation) that must be implemented with distributed functionality in network-centric environments. In particular, these current technology gaps include:

- Ontology consistency, to enable automated machine collaboration across communities of interest;
- Information services, to enable management of information content and quality;
- Automated sensor resource management, coupled to dynamic tactical needs and military operational needs;
- Distributed, heterogeneous, real-time Level 1 data fusion;
- User-defined visualization and automation for decision support; and
- Enterprise monitoring and control, to give the user feedback concerning information processes in terms of performance, expected latency, flow, and quality.

***Recommendation.*** Based upon the findings presented above and on the issues described in this section, the committee recommends the following:

- **Recommendation** for ONR: Monitor technology availability and, as appropriate, invest to sustain investigations that:
    —Develop technology for distributed real-time processing at heterogeneous fusion;
    —Develop resource allocation driven by current operational situation understanding; and
    —Identify and supplement information services that assure consistent information management processes across the network enterprise.

### 6.2.3 Situational Awareness and Understanding

#### 6.2.3.1 Situational Awareness Overview

Automated techniques for achieving situational and threat awareness (often referred to as Level 2 and Level 3 fusion, respectively) are needed to distill the volume of COP-like information expected to become available in network-centric environments for specific user needs. Situational and threat awareness require human reasoning about evidence of battlespace and related activities in order to understand force relationships, interpret activity significance, and anticipate adversary intent. COP-based information is one element of such a reasoning process, but additional contextual information from various subject-matter experts and other knowledge sources is required. Further, machine reasoning about object and events (location, kinematics, identification) in the context of relevant information (e.g., environment, doctrine) is currently infeasible. This is particularly true for battlespace problems of military scale (spatial-temporal dimensions; large numbers of objects, events, activities), and uncertain data. Current capabilities are human-based or very small scale. Available quantitative methods have implicit context and therefore have limited application, particularly in network-centric environments.

### 6.2.3.2 Situational Awareness Challenges

As noted above, machine-based capabilities for providing situational or threat awareness are currently not available. Solving this deficiency in automated (or even computationally aided) situational awareness will require machine reasoning capability to aggregate COP-supplied information about the battlespace along with relevant knowledge of adversary forces, to establish relationships among objects, events, and the environment. Such relationships are key to understanding situations (e.g., red force dispositions, blue force vulnerability) and threats (e.g., red force options, path of intended movement).

Contributing technologies deemed necessary for progress in this area include these:

- *Inference engines*—probabilistic, logical, and so on, for bounded spatial, temporal, abstract properties;
- *Knowledge management*—knowledge bases, sources, acquisition, authoring, and validation tools, probabilistic and uncertainty representation;
- *Large-scale relational and control frameworks*;
- *Human–machine collaboration*—interactive hypothesis management; and
- *Cognitive modeling.*

Goal capabilities include these:

- Automated consistent understanding of situations,
- Automated consistent understanding of adversary intent and threats,
- Adversary intent analysis,
- Anticipation of possible battlespace futures,
- Data mining,
- Information discovery,
- Automatic (or aided) target recognition,
- Activity pattern recognition, and
- Dynamic "what if" analysis.

### 6.2.3.3 Situational Awareness Science and Technology Perspectives

Significant advances in S&T will be required to achieve aided, or automated, situational and threat awareness. The technologies deemed relevant are technologically immature, and the commercial resources that might be leveraged have embedded context that is inconsistent with military operations.

***Findings.*** Technology to provide automated situational and threat awareness is currently not available. In particular, these current technology gaps include:

- Contextual reasoning regarding problems having scale and uncertainty of battlespace issues,
- Knowledge bases and tools to capture and represent diverse battlespace expertise,
- Interactive human–machine hypothesis management, and
- Visualization and cognitive interfaces.

***Recommendation.*** Based upon the findings presented above and on the issues described in this section, the committee recommends the following:

- **Recommendation** for ONR: Monitor technology availability and, as appropriate, invest to sustain investigations that:
   —Advance inferencing techniques necessary to relate objects and events to their environment and to units, activities, and behaviors,
   —Develop a relational and control framework for managing a broad range of knowledge representations, hypotheses, assertions, and so on,
   —Develop automated techniques for information capture, representation, authoring, and validation, and
   —Integrate human and machine capabilities for hypothesis management—balancing machine capability for handling numerical-scale problems with human ability for intuition.

### 6.2.4 Information Assurance

#### 6.2.4.1 Information Assurance Overview

Information is derived from integrating and interpreting data from multiple sources including sensors as well as software and human agents. Information assurance involves the availability, reliability, security, and trustworthiness of this information. The challenge of quantifying the assurance of information in FORCEnet is particularly challenging, since FORCEnet will be composed of multiple heterogeneous systems, often of independent design and different operational origin.

Information assurance is provided by the communications and the collaboration levels defined in the Technology Architecture for FORCEnet. The communications/network core provides basic transportation for the information as well as having responsibility for ensuring its availability and reliability and for the responsiveness of its delivery. The collaboration level provides for information sharing and must ensure the interoperability of the information sources as well as providing indications of the security and pedigree or trustworthiness of this information. The following subsections address information assurance issues with each of these two levels.

### 6.2.4.2 Information Assurance Challenges

***Communications/Network Core.*** While there are mathematical definitions for the long-term availability (i.e., the probability at any instant in time that the resource is usable) and reliability (i.e., the probability that a resource that was usable at time $t = 0$ is still usable at time $t = T$), these measures are difficult to apply to complex systems. Often it is difficult to even define when a resource is usable. For example, while we would consider the telephone system unusable if no one could place a call, would the system be considered usable if half of the people could not place a call? What if only one person cannot place a call? Your definition might depend on whether you were that one person. Historically it has been very difficult to even define the conditions that indicate when a complex system is meeting its functionality. This is especially true when the system may be dynamically composed and reconfigured.

The first goal must be to define metrics for the reliability, availability, and robustness of FORCEnet. These metrics must be measurable with reasonable effort and must not require an omnipresent view of the entire system. The measurements will be composed of predeployment as well as operational assessments. Predeployment assessment is often done via benchmarks. Benchmarks are usually self-contained programs that utilize resources in a synthetically derived manner that mimics real application behavior. Benchmarks are composed of a test to stress a FORCEnet property and a built-in evaluation methodology that reports a quantity that is relatable to the property being exercised. Benchmarks are repeatable, allowing comparisons across systems as well as providing a means to measure improvement in a single system.

Benchmarks exist for performance measuring and specific aspects of software robustness testing. Benchmarks are one way to evaluate COTS components as well as to monitor end-to-end communications and network capabilities. The synthetic workloads that often provide the background stress for benchmarks can also be used to inject workload during operational exercises to see how doctrines work when network resources are strained.

As an example of the new generation of benchmarks, consider Ballista. Since COTS and legacy software will be used in FORCEnet to reduce development time and cost, an automated means for their evaluation is required. COTS software is typically tested only for correct functionality, not for graceful handling of exceptions. Yet studies have indicated that more than half of software defects and system failures may be attributed to problems with the handling of exceptions. Even mission-critical legacy software may not be robust to exceptions that were not expected to occur in the original application (this was a primary cause of the loss of the Ariane 5 rocket's initial flight in June 1996). The Ballista automated robustness-testing methodology characterizes the exception-handling effectiveness of software modules by making calls with exceptional parameters and monitoring the results.[5] Ballista only requires the syntactical definition of the call and a list of each call's argument types. In one study, each of 15 different operating systems' robustness was measured automatically by testing up to 233 POSIX functions and system calls with exceptional parameter values.[6] The study identified repeatable ways to crash these commercially available operating systems with a single call, ways to cause task hangs within operating system code, ways to cause task core dumps within operating system code, failures to implement defined POSIX functionality for exceptional conditions, and false indications of successful completion in response to exceptional input parameter values. While one would expect commercial operating systems to be highly robust, any of these behaviors could be fatal to FORCEnet if the error occurred naturally, and especially if this approach were used in a hostile information attack. Overall in this study, only 55 percent to 76 percent of tests performed were handled robustly, depending on the operating system being tested.

***Collaboration.*** The collaboration level provides for sharing of information in a timely manner. The information will be generated on a variety of platforms, each of which may be protected by different security methods. How should the system reflect the security of the origin of the data that went into composing the information? What level of trust can be placed in that information—that is, what is the pedigree of the information? Assessing the trustworthiness of the software systems will pose major challenges. Coalition software will be globally developed in a distributed manner, software agents will move from platform to platform, and software will be upgraded in the field. How will configuration consistency be managed?

Metrics for security and trustworthiness need to be developed. The measures should allow composing values from components into an end-to-end measure that can be appended to information fragments and that not only can be presented in an intuitive way to the user without distracting the user from the primary data, but also that are not overwhelming the user with this auxiliary information. Benchmarks should be developed to evaluate the security and trustworthiness of components and end-to-end systems composed from components. Monitoring techniques need to be developed to automatically detect intrusions and insider threats. Wherever possible, the response to these threats should be automated. On the basis of these data models for systems should be developed that allow predict-

[5]Nathan P. Kropp, Philip J. Koopman, and Daniel P. Siewiorek. 1998. "Automated Robustness Testing of Off-the-Shelf Software Components," *IEEE Proceedings of the Fault Tolerant Computing Symposium '98,* Munich, Germany, June, p. 1.

[6]Philip J. Koopman and J. DeVale. 2000. "The Exception Handling Effectiveness of POSIX Operating Systems," *IEEE Transactions on Software Engineering,* Vol. 26, No. 9, p. 837.

ability of system behavior and allow for balancing between protection and information sharing. A systematic red team activity could help with "testing the testing."

In those cases in which the theory is insufficiently developed, it may be necessary to depend on human monitoring of FORCEnet. As an example, the Computer Emergency Response Team Coordination Center (CERT/CC) receives vulnerability reports, verifies and analyze the reports, provides technical advice, coordinates responses to security incidents, works with other security experts to identify solutions, identifies trends in intruder activity, analyzes software vulnerabilities, and disseminates information to the community.

CERT/CC operates in a feedback cycle. It analyzes flaws in Internet systems, measures the exploitation of those flaws, assists in remediation, and studies intruder-developed code that exploits the flaws. During analysis, a catalog of artifacts and analysis is built; on the basis of this catalog, a capability is developed to predict trends in malicious code development and functionality. CERT/CC provides 24-hour emergency incident response to threats and attacks on the Internet infrastructure, to widespread automated attacks against Internet sites, and to new types of attacks or vulnerabilities. In 2002, CERT/CC reported 82,094 incidents and 4,131 vulnerabilities and processed 204,697 e-mail messages.[7]

### 6.2.4.3 Information Assurance Science and Technology Perspectives

*Findings.* Inadequate technology exists to provide the necessary level of information assurance to support the FORCEnet vision. The network need for sharing information must be balanced with traditional information assurance roles involved in protecting information. This challenge is made more difficult under conditions requiring trusted information exchange across multiple independent levels of security and among coalition partners. In particular, the current technology gaps include these:

• Metrics, automated network analysis, and monitoring of network reliability and security that are capable of scaling to network-centric needs and to the demands of multilevel security and failure prediction;
• Dynamic balancing of protection levels, including policy adaptation, with sharing needed to maintain mission effectiveness;
• Trustworthiness of software systems and associated information in network-centric operations; and
• The ability to conduct intrusion detection and identify insider threats.

---

[7]Current statistics are available at http://www.cert.org. Accessed July 24, 2004.

***Recommendation.*** Based upon the findings presented above and on the issues described in this section, the committee recommends the following:

- **Recommendation** for ONR: Assess FORCEnet applicability of information assurance technology and, to the degree required, sustain investigations seeking to develop:

    —Improved metrics for information assurance,

    —Automated, real-time, network-centric systems analysis to identify and predict information systems failures, and

    —Improved techniques to achieve multiple levels of information security.

### 6.2.5 Modeling and Simulation

#### 6.2.5.1 Modeling and Simulation Overview

A clear appeal for help with respect to modeling, simulation, and analysis was heard throughout the course of this study. Some of the requests require a technical solution and some require a political or social solution. The subsections below discuss both, so as to not lose the flavor of the requests for help.

With respect to network-centric warfare itself and, in particular, with reference to lessons learned from OIF, it was often mentioned to the committee that at times the Navy had no analysis capability for checking out networks and network-dependent systems before they became operational or for really checking out whether or not network-centric warfare was a real possibility with the available or future network communications infrastructure. Upon further inquiry, the committee found that there were not enough people, either civilian or military, with modeling, simulation, and analysis backgrounds. There was an apparent lack of properly trained simulationists readily available when needed. This personnel shortage was exacerbated by the lack of appropriate modeling and simulation toolsets.

#### 6.2.5.2 Modeling and Simulation Challenges

The committee found that there are many modeling and simulation (M&S) tools for fighting the Cold War, red against blue, but there are few available tools for modeling, simulating, and analyzing the types of warfare being pursued now. This is especially true with regard to modeling and simulation of a large-scale, network-centric environment. Accurate modeling and simulation are also needed to support network-centric operations systems engineering and architectures decisions. These large-scale M&S capabilities will be needed to support network-centric operations testing, evaluation, and validation. Few tools really allow "what if" analysis with the asymmetric threat, and none allow modeling, simulation, and analysis of tightly constrained urban environments with guerilla fighters intermixed with coalition and native populations. The military Services have many old-style Lanchester-equation types of Cold War simulations and few ready-to-go, agent-based simulations that allow such explorations.

The Navy needs to create a career path for modeling, simulation, and analysis in its officer corps similar to the recently created information professional career path. Billets for such expertise need to be created in the fleet so that expertise is readily available and so that modeling, simulation, and analysis become part of its tactical operations. The Navy needs analysts as well as modeling and simulation technologists in the fleet. The modeling and simulation technologists need to help build the modeling and simulation systems required now and in the future. Just deploying analysts will not fill this gap in capability.

The Navy needs to invest in the R&D required to build next-generation combat modeling systems. At the present time, the Navy cannot simulate the types of wars being conducted. The Navy cannot get simulation and analysis done in such a way that it can properly know how to deploy FORCEnet. The Navy does not have the technology in hand, nor is it imminently attainable. A major investment in the broader scope of next-generation modeling, simulation, and analysis is essential to properly understand FORCEnet and its impact on the future of warfare.

### 6.2.5.3 Modeling and Simulation Science and Technology Perspectives

*Findings.* The present state of modeling and simulation does not scale to FORCEnet needs. In particular, the current technology gaps include these:

  • Scaling of models and simulations to large numbers of sensors, platforms, and users;
  • Systems engineering, including means to check out large-scale network-centric systems prior to deployment (i.e., the ability to model systems life-cycle design and testing and model validation); and
  • Robust "what if?" analysis to support trade-offs among network-centric system configurations as a function of mission or threat environment, performance, and reliability.

*Recommendation.* Based upon the findings presented above and on the issues described in this section, the committee recommends the following:

  • **Recommendation** for ONR: Sustain investigation in the following areas in coordination with other relevant R&D activities across the DOD, industry, academia, and the commercial sector:

        —Modeling and simulation to support large-scale systems engineering, and
        —Adversarial analysis models and simulations.

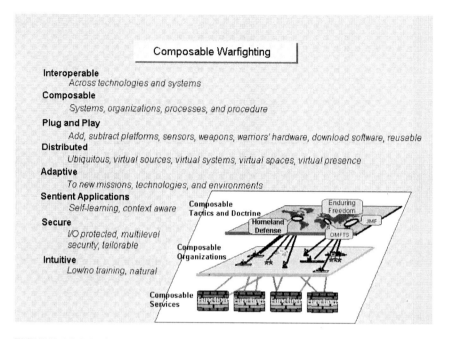

FIGURE 6.5 FORCEnet support of dynamic force composition. SOURCE: Space and Naval Warfare Systems Command. 2003. FORCEnet Government Reference Architecture, Version 1.0, April.

## 6.2.6 Dynamic Composability and Collaboration

### 6.2.6.1 Composability Overview

Central to the vision of FORCEnet is the ability to "compose" system-like capabilities from components of the enterprise in order to achieve effective warfighting readiness in response to dynamically changing operational situations. The concept of FORCEnet composability is discussed at some length in the FORCEnet Government Reference Architecture (GRA) Vision, Version 1.0.[8]

The notion of composability has a number of implications for warfighting and warfighting systems, as captured in Figure 6.5, taken from the GRA. The figure also illustrates the potential flexibility that FORCEnet enables in supporting adaptation in the following separate domains:

• *Technology*—by accessing available services to reconfigure technical functions,

---

[8]Published April 2, 2003.

- *Organizations*—adopting new functions to change organizational roles and relationships, and
- *Tactics, training, and procedures*—assigning new functionality and relationships to adjust processes and procedures to meet new operational challenges.

### 6.2.6.2 Composability Challenges

To achieve the vision of composability, there is a need for information management technology (as discussed earlier) to meet mission goals, along with an enterprise-wide oversight process to assure that aggregated mission goals satisfy campaign outcomes. The former need supports the composition of assigned assets to satisfy individual mission threads. The latter might be viewed as a campaign-level control function for the dynamic balancing of available resources to mission sets, as a function of costs, risks, and expected value. The GRA recognizes this need for layered functionality and control, and includes discussion of the implications of mission-oriented composability.

For mission composability, it is assumed that the architectural framework, open standards, and protocols that accompany enterprise services (and potentially information services) will provide the characterization needed to enable composability. The issue is illustrated in Figure 6.6, from the GRA. The figure shows a variation of the layered functional view discussed earlier, with each layer partitioned into building blocks that may be logically connected in any number of ways to achieve desired mission capabilities.

Composability issues are being further explored in the SPAWAR Systems Center, San Diego's "Command Center of the Future," with experimentation addressing composable engineering issues. Under the slogan "Concept to Capability," the activity is investigating issues relevant to the FORCEnet composable vision and realization of the capability. It is doing so by illustrating the concept potential in an interactive simulation environment, implementing selected components, and identifying benefits (at least qualitatively), along with enabling technologies (middleware and architectural constructs), and decision-support tools. The use of this facility in continuing experimentation provides a useful venue for gaining significant insight into composability issues that are currently not well understood:

- Identification and characterization of needed building blocks,
- Tagging of data, and
- Automated (composition) process management.

The idea of extending this laboratory to be part of a distributed development capability, called FORCEnet Composable Environment, has apparently been discussed as a project of the Virtual SYSCOM program.

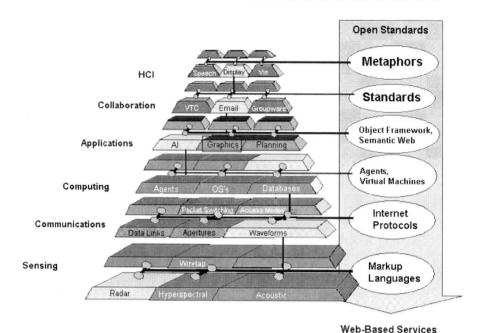

FIGURE 6.6 Requirements for "Composable" C4ISR. NOTES: HCI, human–computer; VTC, video teleconference call; API, application program interface; AI, artificial intelligence; OS, operating system. SOURCE: Space and Naval Warfare Systems Command. 2003. FORCEnet Government Reference Architecture, Version 1.0, April.

### 6.2.6.3 Composability Science and Technology Perspectives

Incorporating the notion expressed above, of campaign-level control, into these notions of mission composability will be a difficult challenge. Very little work has been done to address the complexity of managing mission composability in a way which assures that integrated, dynamic mission goals will achieve desired campaign outcomes. An aspect of this issue is real-time feedback to decision makers about the state of enterprise readiness. Given the complexity of a FORCEnet enterprise, how does a force commander know that the "composition of the moment" will satisfy planned objectives? A readiness monitor that confirms the state of the enterprise—core plus communities of interest—must be contemplated.

***Findings.*** Today's technology does not support dynamic composability "on the fly." A number of essential elements must be addressed to achieve the FORCEnet vision of mission-composable capability while maintaining campaign-level control, required for the coordination of forces. In particular, the current technology gaps include:

• The complexity of managing mission composability in a way that assures integrated resource allocation to meet dynamic mission goals and achieve desired campaign outcomes;
• A readiness monitor that confirms the state of the entire FORCEnet enterprise (core plus communities of interest, across all enterprise layers) in any given configuration, for all users;
• Manpower and training programs to teach and utilize composed functionality; and
• Tools to support automated means of facilitating collaboration between people and/or machines and to include planning or replanning functions.

***Recommendation.*** Based upon the findings presented above and on the issues described in this section, the committee recommends the following:

• **Recommendation** for ONR in concert with the CFFC, NETWARCOM, NWDC, and appropriate laboratory organizations: Commit to a long-term, co-evolutionary process, which involves laboratory and field experiments, to evolve the required technical components with naval tactics and procedures and personnel implications. Thus, the committee recommends that:
—ONR sustain investigation of complex resource management (allocation and coordination) issues.
—ONR sustain investigation of automated collaboration tools necessary to facilitate interactions and problem solving between humans, between machines, and between humans and machines (the effort should also address issues associated with the variable reliability of the naval communications).

### 6.2.7 Support of Disadvantaged User-Personnel, Platforms, and Sensors

#### 6.2.7.1 Disadvantaged User Overview

Dismounted troops or those in small vehicles must be provided network connectivity and situational awareness. This needed capability places a premium on the power, weight, and size of the electronic system used by the troops. A particularly challenging problem is providing situational awareness to disadvantaged users while not distracting them from carrying out their primary missions.

Batteries have been a particularly serious problem in Operating Enduring Freedom and OIF. Energy densities need to be increased as well as battery lifetime and rechargeability. Fuel cell technology has made promising advances in the area of powering small personal electronics. Nanotechnology may also provide new approaches to improve energy density—for example, through the application of carbon nanotubes to battery elements.

Reducing the weight and energy consumption of electronics would also have a high payoff. There is no reason why the weight and consumption of receive-only communications equipment should not be able to be reduced substantially. Nanotechnology approaches may have payoff here, as well as in self-generated power. Nanoelectronics could also help reduce overall power consumption by as much as a factor of 10.

### 6.2.7.2 Disadvantaged-User Science and Technology Perspectives

*Findings.* The state of technology to support disadvantaged users (for example, small boats, dismounted marines) is deficient. Depending on operational conditions, unique needs may exist for communications, information representation, and human–machine interfaces. Beyond issues cited in other sections, the current technology gaps include:

• The human–machine interface—today's handheld displays are difficult to read and distracting, and head gear is bulky and also distracting;
• Custom representation of information to meet difficult operating conditions;
• The size and weight of antenna apertures too large for routine use by disadvantaged (especially dismounted) users; and
• Power sources too heavy and bulky for rapid mobile use by individuals.

*Recommendation.* Based upon the findings presented above and on the issues described in this section, the committee recommends the following:

• **Recommendation** for ONR, in coordination with DARPA and the Army: Sustain investigations that seek to provide:
    —Minimum essential situational awareness for dismounted troops by means of technology that is least distracting for the users, and
    —Lightweight, high-density power sources and improvements in power consumption in coordination with DARPA and the Army.

### 6.2.8 Persistent Intelligence, Surveillance, and Reconnaissance

#### 6.2.8.1 Persistent Intelligence, Surveillance, and Reconnaissance Overview

The concept of network-centric operations and its implications for composability require not just access to data from a range of national, theater, and tactical sensors but more complete integration of sensing components into the network framework. For FORCEnet, such integration implies (1) adaptive control (optimization) of ISR operations to satisfy dynamic changes in battlespace priorities and in areas and activities of interest; (2) coordination of sensing functions and geometries with enterprise goals; and (3) robust sensing functionality to provide observations in the appropriate modality over the wide areas of concern, at time rates appropriate to mission functions, and in forms suitable for automated processing and aggregation.

Sensing components in this context refer to the broad set of sensing options ranging from current (or equivalent) sensing capabilities available for national, theater, or tactical operations, to emerging concepts for novel sensing modalities, as well as innovative new concepts for robust distributed and autonomous sensing. In the latter case, studies are being conducted to assess the feasibility and efficacy of large distributed fields of very low cost, self-organizing sensors that are capable of monitoring large areas and of being configured to exfiltrate significant indications and observations to the larger network.

### 6.2.8.2 Persistent Intelligence, Surveillance, and Reconnaissance Challenges

Recognizing that the supply of sensors is unlikely to meet the separate demands for coverage of a multiplicity of users, the network challenge is to manage sensor coverage in a way that best serves command priorities and mission needs. This need to maximize and adjust the value of collected information to the enterprise carries the strong implication that for network-centric operations, the burden of assured sensor coverage—for both offensive and defensive purposes— must shift from a narrowly focused, personnel-intensive, platform-centric perspective to a multimission, multifunction, automated network of sensors capable of providing adaptive coverage for all enterprise users.

Whatever the combination of deployed sensors, the notion of pervasive and persistent ISR implies a network ability to monitor the state of sensing at any given time and to respond dynamically to changing needs in any region of the battlespace volume. Broadly stated, the FORCEnet need for persistent and pervasive ISR includes the following:

• *Persistent sensing*—refers to the ability to maintain coverage and to update observations at rates suitable to the intelligence, surveillance, reconnaissance, or targeting function required by the mission and driven by the expected target behavior. It is worth emphasizing that observation time and revisit requirements will vary markedly as a function of sensor type, target activity, background conditions, and, most importantly, mission function. Virtual continuous observation is seldom necessary or appropriate for most sensors. Persistence in this sense implies an effective sampling rate that can be adjusted to suit sensor modality, battlespace conditions, and mission needs. This notion of adaptive sensor collection rhythm to match mission functions suggests that sensing goals might be interleaved in network environments to maximize sensor utility.

• *Pervasive sensing*—refers to the ability to achieve and maintain observability within the coverage envelope in modalities appropriate to the activity and environment of interest. For difficult targets or background conditions that limit target observability, it is understood that multiple sensors, or new sensing modalities, may be required in order to achieve the required performance in detection, tracking, or identification. It should also be noted that even in circumstances in which the user's observation needs can be satisfied by a single preferred sensing means, in many cases a combination of sensors, properly deployed and fused, can provide an equivalent answer. This ability of the enterprise to implement alternative satisfaction methods complements the notion, expressed above, of adaptive sensor collection rhythm, and it presents both an opportunity (to reduce the disparity between supply and demand) and a challenge (to perform dynamic reallocation) for network technology.

• *Command and control of platforms and sensors to meet dynamic operational and tactical needs*—refers to the need in network-centric operations to dynamically reallocate resources in order to meet changes in understanding and priorities of the battlespace situation. In enterprise environments, given constrained resources, a means to continually assess the allocation of resources to organization, mission function, area, activity, and time will be needed.

• *Dynamic planning and replanning of sensor modality and coverage*—refers to the needs for the automated allocation of sensing resources to tasks and for the automated routing and scheduling of sensor activities, including geometry, time, and mode. In order to be responsive to target behavior, an anticipatory (predictive) strategy will be needed to overcome latencies inherent in replanning and redirection of sensors and platforms. Automation will be essential in order to maintain efficient utilization of sensor assets by the enterprise and to enable coordinated (and often synchronized) use of sensors across the network to achieve desired operational effects.

• *Distributed, autonomous sensor networks*—has particular relevance to unattended sensors deployed either as autonomous vehicles (e.g., unmanned air or ground vehicles) or distributed in large numbers over large areas or volumes to provide some alerting or monitoring function. In addition to the need for the development of device and packaging technology, such vehicles or distributed fields of sensors will likely require some ability to do the following: self-organize in response to their physical distribution and environment, self-monitor and adapt in response to degradation or loss of sensing elements, and adaptively control processing or communications functionality within predetermined limits in response to detected events or command instructions.

A final comment on the role of humans in persistent and pervasive ISR is appropriate. As suggested above, the implication of network-centric operations is that a high degree of automation will be required to perform a large number of functions (some relatively low level) now performed by humans (e.g., platform and sensor routing, sensor scheduling, mode control, data tagging, analysis). The dimension envisioned for FORCEnet, together with the requirements for parallel activities, speed of response, and rapid adjudication of needs across multiple nodes and users, suggests that humans will be inappropriate for such roles in a fully evolved network-centric environment. This argument is completely apart from, but consistent with, the trend in current operational and support environments to reduce staffing and subject-matter experts in every area. Humans will need to interact with automated capabilities to provide supervision and confirmation of critical products. The challenge for technology and the coevolutionary process, discussed in Chapter 2, will be to identify the appropriate degree of automation and the appropriate mechanisms to interface automated products with humans in order to achieve human–machine collaboration in network-centric operations.

### 6.2.8.3 Persistent Intelligence, Surveillance, and Reconnaissance Science and Technology Challenges

*Findings.* For network-centric operations, traditional intelligence, surveillance, and reconnaissance sensing (national and theater, platform-based coverage) will need to be augmented by organic tactical sensors for the responsive coverage of areas that are difficult to monitor, or for which access is denied. In particular, the current technology gaps include these:

• Automation for the coordinated usage of multiple sensors, adaptive sensor control, and more robust sensing modalities;
• Automation to drastically reduce personnel requirements and to reverse the ratio of humans per sensor from a positive number to a fractional number—this will become especially important with the likely proliferation of small sensors for wide-area coverage of difficult areas; and
• Small, networked sensors for wide-area, inexpensive alerting in difficult or denied areas.

*Recommendation.* Based upon the findings presented above and on the issues described in this section, the committee recommends the following:

• **Recommendation** for ONR: Monitor technology availability and, as appropriate, invest to sustain investigations in:
   —Networked sensor technology for wide-area alerting of asymmetric targets or activity,
   —Automated sensor management for adjudicating sensing needs across mission goals and for sensing responsiveness to dynamic battlespace needs, and

—Machine-to-machine collaboration for remote operations.

### 6.2.9  Summary of Functional Capabilities and Challenges

The issues discussed throughout Section 6.3 cover the breadth of technical functionality required to realize a fully functioning FnII capability. The technical challenges to achieving the FORCEnet goal are considerable and will require significant R&D investment and innovation by the naval forces. A summary of the challenges perceived at this time is provided in Table 6.1. The table offers a high-level overview of the most significant S&T issues involved in each of the functional capabilities discussed in this chapter. The table includes a column ("Level of Assessment/Maturity") that provides the committee's estimate regarding the degree to which the ONR Future Naval Capabilities (FNC) program (primarily 6.3 (advanced technology development) funding) will address perceived challenges. The challenges cited in Table 6.1 that are properly the focus of 6.1 (basic research) or 6.2 (applied research) effort may be addressed under the ONR's Discovery and Invention (D&I) program. D&I plans were not available in time for committee consideration.

Table 6.1 serves to emphasize the range and difficulty of the technical challenges facing the development of network-centric capability, along with the varying levels of available technology maturity. Achieving the vision of network-centricity will not be easy.

### 6.3  ENABLING TECHNOLOGIES RESULTING FROM THE GLOBAL INFORMATION GRID

#### 6.3.1  FORCEnet Implications of the Global Information Grid

The GIG initiative is addressing some major S&T issues that must be overcome to ensure its success. GIG activities, directed by OSD, provide significant motivation for transformation of the Services to network-centric operations. While the OSD provides guidance, direction, and some investment to the Army, Navy, Air Force, and joint communities, it is understood that realization of the GIG in support of combat capability depends on developments and experimentation within, and across, each of the Services and agencies.

The Navy should examine all S&T programs that are relevant to the GIG and determine if these programs will provide robust solutions for the Navy. For example, the GIG is working to establish high-bandwidth communications for mobile users, but it may not address all of the problems that need to be solved to ensure successful naval operations. The GIG is operated under the assumption that there will be continuous connectivity, which has major impact on integrated topside RF architectures. The Navy must have mobile, adaptive networks to support ships at sea and disadvantaged users afloat and ashore. There are similar

TABLE 6.1  Science and Technology (S&T) Challenges and Levels of
Maturity for FORCEnet Functional Capabilities

| Capability | Challenges | Level of Assessment/Maturity | Office of Naval Research S&T |
|---|---|---|---|
| Wideband communications | Linking of data rate and antenna agility. | Mature except for the optical regime. | P |
| | Automated link and network monitoring and control. | Requires metrics and algorithms for maximizing mission effectiveness. | N |
| | Protocols to deal with mobility, disruptions, and security. | Complex trade-off solution space is not well understood; requires modeling and experiments. | P |
| Information management | Information services to maintain information content and quality. | Technically feasible—requires focus on information-centric coordination and discipline. | Y |
| | Ontology consistency for machine collaboration with user supervision. | Coordination needed within communities of interest to coevolve technology with procedures and tactics; automated decomposition of information needs to tasks is a 6.1/6.2 issue. | P |
| | Automated sensor and resource management, coupled to dynamic needs. | Automated allocation of tasks to assets in high-demand conditions is a 6.3 problem. | P |
| | Distributed, multisource, real-time Level 1 data fusion. | Distributed, real-time processing currently infeasible—6.2 development is needed; architectural options are available for interim solution. | N |
| | User-defined visualization; appropriate degree of automation. | Information packaging and representation are near-term feasible; cognitive research is required for longer-term solutions. | Y |
| Situation understanding | Contextual reasoning regarding problems having scale, and uncertainty of battlespace issues. | Inferencing techniques are available—all of small scale with bounded constraints; development and experimentation are needed to extend performance bounds. | Y |
| | Knowledge bases and tools for diverse battlespace expertise. | Numerous R&D issues exist in capture, representation, authoring, and validation. | P |

*continues*

TABLE 6.1 Continued

| Capability | Challenges | Level of Assessment/Maturity | Office of Naval Research S&T |
|---|---|---|---|
| | Interactive human–machine hypothesis management. | Significant R&D is required to balance machine capability for numerical scale, with human ability for intuition. | P |
| Information assurance (IA) | Automated monitoring and analysis of system IA. | Development of improved metrics and monitoring techniques is needed. | N |
| | Collaboration exacerbates Coalition information assurance issues. | Improved means of assuring information integrity are needed. | P |
| | Trustworthiness of network-centric operations (NCO) systems is inadequate. | Network monitors and intrusion-detection capability are needed. | N |
| Modeling and simulation (M&S) | No means to check out systems prior to deployment in Operation Iraqi Freedom. | M&S for large-scale system development is inadequate. | N |
| | No robust "what if" analysis. | M&S funding is needed for stochastic, adaptive accurate analysis of adversaries and the environment. | N |
| Composability/ collaboration | Monitoring and control of enterprise information configuration and readiness. | Requires development of information-centric metrics and modeling techniques—6.2 effort. | N |
| | Automated management of network resources; automated collaboration of machines and users is inadequate for NCO. | Large-scale optimization techniques are available—application to problems of military scale and dimensionality are not understood; 6.2 focus is needed. | N |
| Disadvantaged user | Automated situational awareness and human–machine interface are inadequate. | Deployed situational awareness with easy human–machine interface needs improvement. | P |
| | Processing power needs to be improved; processor size and weight are issues. | High-performance processing and displays must be reduced in size and power. | N |

TABLE 6.1 Continued

| Capability | Challenges | Level of Assessment/Maturity | Office of Naval Research S&T |
|---|---|---|---|
| | Energy generation and consumption must be improved. | Power sources are too heavy. | P |
| Persistent intelligence, surveillance, and reconnaissance | Sensing components and techniques for difficult targets and environments. | Identification and development of novel sensors and nontraditional sensor usage are needed. | Y |
| | Persistent sensing—ability to allocate coverage and revisit rate in accord with area and activity of interest (A/AOI), and mission purpose. | Making automated sensor/network management (see above) responsive to difficult mission needs and A/AOIs is a 6.2 issue. | N |
| | Pervasive sensing—ability to maintain type of coverage appropriate to the activity and environment of concern. | Derivation and allocation of sensing needs from phenomenology are a poorly understood problem. | P |
| | Anticipatory capability derived from target behavior and environmental context; dynamic reallocation of sensors by location and modality. | Dynamic, automated platform/ sensor replanning of routes, schedules, and tasking responsive to target behavior and mission priorities is a 6.2 issue. | N |
| | Autonomous sensor networks for alerting and monitoring of large and difficult areas and activities. | Sensing, networking, self-adaption, and power conservation are all embryonic technologies. | Y |

NOTES: Y = S&T planned; N = S&T not evident in plans; P = partial scope of S&T addressed. Department of Defense budget activities: 6.1 = basic research; 6.2 = applied research; 6.3 = advanced technology development. A list of acronyms is provided in Appendix C.

issues with optical communications in the marine layer for surface ships. Underwater vehicles have very restricted reliable communications, especially at speed and depth. Similarly, disadvantaged user-personnel, platforms, and sensors do not have great flexibility with regard to antenna apertures, and they have need for high-density, low-weight power sources. There are issues with low overhead information assurance for the disadvantaged user. Automated collaborative technologies and automated composability do not work well in an unreliable communications environment.

The Navy also needs a comprehensive, shared situation understanding that includes the identification and tracking of surface and subsurface vehicles. Information management is very important, to ensure that small disadvantaged users and platforms receive the proper information in a timely fashion without causing information overload. All of these issues need to be addressed in the Navy S&T to support FORCEnet.

The ONR's FY 2006 S&T plan, responding to N70's statement of FORCEnet gaps, explicitly addresses GIG leveraging in a number of places. Specific plans include research to explore naval-unique concerns in communications, apertures, information management, and decision support (community-of-interest technologies), and information assurance. In the area of enterprise services, ongoing efforts have developed Services Oriented Architecture (SOA) capabilities, which are contributing to the DISA's Network-Centric Capabilities Pilot Program as well as to the Horizontal Fusion effort of the ASD(NII). Planned efforts will continue to explore issues related to SOA technology and utility. A proposed Advanced Concept Technology Demonstration specifically addresses a near-term enterprise services demonstration enabled via NCES eXtensible Tactical C4I Framework (XTCF) technology.

Participation by naval PEOs includes the contribution of the PEO C4I & Space to DISA's NCES development and to migration of the GCCS (with its centralized architecture) to Joint Command and Control (with its Services-oriented architecture). The open architecture program of the PEO IWS is participating in the ASD(NII)'s Horizontal Fusion demonstration. Within the systems commands, the Office of the Chief Engineer (SPAWAR 05) has partnered with the Air Force's Electronic Systems Command to evolve enterprise architecture issues common to FORCEnet and to the Air Force Command and Control Enterprise Reference Architecture. Such coordination will help assure consistency as experience with network-centric operations accumulates.

## 6.4  SCIENCE AND TECHNOLOGY PROGRAM OF THE OFFICE OF NAVAL RESEARCH

Science and technology relevant to FORCEnet has been addressed by ONR under its Future Naval Capabilities programs, primarily Knowledge Superiority and Assurance and to some extent under Autonomous Operations and Fleet Force Protection. Enabling capabilities as defined through FY 2004 have addressed areas relevant to network-centric operations. The capabilities addressed include these:

- Common, consistent knowledge;
- Distributed, collaborative planning and execution;
- Enterprise-wide integrated information; and

• Dynamically managed, interoperable, high-capacity connectivity, shared apertures, networking, interoperability.

Products of these efforts to date include:

• *XTCF*—a framework for network-centric information management for naval and joint forces to operate in a GIG-defined environment; this effort will provide an early instantiation of enterprise services capability for NCES;
• *Analytical Support Architecture*—a tool for the automated intelligence assessment of enemy air defense;
• *Environmental Visualization*—fused, interpreted, analyzed environmental information disseminated within less than 1 hour of collection;
• *Rapid Maritime Identification and Tracking System*—near-real-time biometric data for maritime special operations forces, to improve their ability to find people and take action;
• *Multinational Virtual Operations Capability*—near-real-time joint force and coalition force exchange of tactical and operational information.

During the course of FY 2004, the ONR program was to be restructured around a new description of enabling capabilities more directly related to warfighting gaps identified by OPNAV. The restructuring of enabling capabilities and the determination of technology content are in progress as this report is being written. For FORCEnet, the organization of the program is expected to be consistent with the MCPs defined by the Naval Transformation Roadmap:[9] Networks, ISR, and Common Operational and Tactical Picture (COTP), plus essential supporting technology primarily in information assurance.

ONR program content is expected to be driven by the following set of shortfalls, identified by N71:[10]

• Vulnerable links,
• Saturated links,
• Insufficient communications and network structure,
• Inadequate network defense in depth,
• Limited coalition interoperability,
• Inadequate sensor strategy,
• Limited undersea picture,

---

[9]Gordon England, Secretary of the Navy; ADM Vern Clark, USN, Chief of Naval Operations; and Gen James L. Jones, USMC, Commandant of the Marine Corps. 2002. *Naval Transformation Roadmap: Power and Access . . . From the Sea*, Department of the Navy, Washington, D.C.

[10]RADM Thomas E. Zelibor, USN, Space, Information Warfare, Command and Control (N61), "FORCEnet POM 06 Process," presentation to the National Defense Industrial Association, San Diego Chapter, San Diego, California, October 23.

- Lack of common maritime picture, and
- Multiple combat identification and BFT solutions.

Technology issues under consideration within each of these areas are summarized below:

- Networks;
  — Multibeam and multiband apertures;
  — High-data-rate communications on the move—including relay, router, quality of service, and network management issues;
  — Undersea and marine layer communications;
  — ISR;
  — Smart sensor networks;
  — Ship's missile defense;
  — UAV-borne robust surveillance;
  — COTP;
  — COTP integration and dissemination to all users;
  — Near-real-time fusion of intelligence and tactical data;
  — Spatial-temporal registration of multisensor data (imaging and non-imaging);
  — Intelligent and adaptive sensor management;
  — Automated situational and threat awareness; and
  — User-tailorable information feeds and displays.
- Crosscutting/Leveraging;
  — Information assurance;
  — Real-time, multicombatant command, engagement planning and control; and
  — Real-time deconfliction of targeting information.

The ONR program planning effort addresses the FY 2005 to FY 2011 time frame. Trade-offs involving program priorities, resource allocation, and start or delivery timing have not been resolved as of this writing.

Table 6.2 provides a listing of program issues and capabilities under consideration by ONR's Knowledge Superiority and Assurance FNC as of the second quarter of FY 2004. This FNC represents the principal, but not sole, investment in FORCEnet development for ONR. The table does not reflect probable Discovery and Invention investment relevant to FORCEnet.

Overall program planning at ONR indicates a strong commitment to the development of relevant network-centric technology, with a broad range of the previously identified shortfalls being addressed. The ONR plans do appear to address a substantial subset of the technologies needed for near-term experimentation with network-centric capability. Given the difficulty of the network-centric

TABLE 6.2  Summary of Knowledge Superiority and Assurance Program Plans of the Office of Naval Research (ONR) as of January 2004 (Preliminary for Period FY 2005 to FY 2011)

| Planned ONR Science and Technology Program | Capability |
|---|---|
| **Common Operational and Tactical Picture (COTP)** | |
| Joint Real-Time Coordinated Engagement | Real-time multicombatant commander coordinated engagement planning and control. |
| Automated Situation and Threat Assessment | Automated production of actionable information for battlespace understanding and prediction of future battlespace activity. |
| Actionable Information from Multiple Intelligence Sources in the GIG-ES Environment | Fused intelligence and cryptologic information: delivered faster, with better quality for command and control of weapons systems. |
| Improved Maritime COTP in the GIG-ES Environment | Accurate, timely surface and undersea information, with reduced false-alarm rate and views tailorable to user need. |
| Decision Support for Dynamic Target Engagement | Significantly decreased time required to engage a pop-up dynamic target. Reduces several key bottlenecks in the decision process for strike warfare. |
| **Networks** | |
| Communications in S-Ku multifunction aperture | Advanced multifunction system that combines communications with electronic warfare in a single system. Includes line-of-sight and satellite communication capability. |
| Ultrahigh-Frequency (UHF)/L-Band Phased Array Antennas for Aircraft Carrier (CVN) | CVN capability to support up to 20 ultrahigh-frequency and L-band communication links, with significantly fewer antennas topside. |
| High-Altitude Airborne Relay | ISR range extension and communications to highly mobile naval forces, reduces dependency on SATCOM, connects to GIG Transformational Communication System (TCS). |
| Expendable Airborne Relay and Router | Large numbers of expendable communications relays and routers over the battlespace, a tactical complement to GIG TCS. |
| Mobile Dynamic Quality-of-Service Enabled Networks | End-to-end mobility using all network links for maximum bandwidth efficiency, with expedited service for high-priority traffic flows. |
| Integrated and Autonomous Network Management | Automated monitoring, configuring, and troubleshooting of networks without human action, and design and implementation of networks using models and simulations to predict performance prior to operations. |
| Environmentally Adaptive Networks | Active monitoring of environments, proactive prediction, and optimization of performance parameters. |

*continues*

TABLE 6.2  Continued

| Planned ONR Science and Technology Program | Capability |
|---|---|
| Optical Communications Through the Marine Layer | Reliable line-of-sight networking capability among surface, ground, and air platforms. |
| Undersea Optical Communications | Reliable undersea networking capability among subsurface platforms and sensor nodes, resulting in improved undersea situational awareness. |
| **Intelligence, Surveillance, and Reconnaissance (ISR)** | |
| S-Band Missile Defense Radar | Full-volume air surveillance to detect and discriminate ballistic missiles; multiple target tracking, long-range target identification, with high-performance in clutter and countermeasures. |
| Reconfigurable Surveillance Unmanned Air Vehicles | High-resolution imaging of potential threats day and night, through fog, rain, and camouflage. |
| Smart Sensor Networks | Networks with large numbers of unmanned, autonomous sensors to provide persistent, pervasive battlespace ISR. |
| **Information Assurance** | |
| Secure Distributed Collaboration | Secure dissemination of information across multiple joint/coalition boundaries. |
| Network Assessment, Monitoring, and Protection | Protects naval IP networks from network attacks and provides for remote administration of networks in response to attacks. |
| Assured and Trusted Computing | Denies adversaries the ability to corrupt software, data, and information on naval networks, both in storage and during transmission. |

NOTE: Acronyms are defined in Appendix C.

challenge, it is not surprising that differing levels of R&D investment (6.1 to 6.3) are required and that not every issue is being addressed with equal attention.

Evolutionary development implies a phased approach to technology development, and ONR efforts have an implicit phasing supportive of an evolutionary development process. Each phase is modulated by feedback from user experience gained through hands-on experimentation and by refined requirements derived from evolving concepts of operation, tactics, and procedures. Toward this end, ONR has engaged in continuing interaction with the units responsible for requirements (N6/N7) and acquisition (PEOs and systems commands), and with fleet representatives (NETWARCOM, NWDC), to guide its planning activities and to establish success criteria for its technology products. While such a phased approach supported by a coevolutionary or experimentation process has not been made explicit, an informal basis has been established.

ONR program personnel have also been active in maintaining awareness of other Service and agency developments so as to leverage progress and to coordinate experimentation wherever possible. In this regard, interaction with the Air Force Research Laboratory, DARPA, and DISA have been particularly notable.

Beyond the ONR science and technology program, it is understood that technologies relevant to FORCEnet success are being addressed across the spectrum of DOD, industry, and academic activities. General awareness of these activities resides within the committee, and their potential contribution to FORCEnet capability was factored in to the discussion above. No attempt was made to perform a more complete survey and assessment. As indicated above, ONR program officers and their trusted advisers do attempt to plan S&T activities with awareness of the community state of the art, in order to take advantage of emerging technology, to leverage opportunities, and to avoid redundant effort. The committee suggests that the Navy investigate relevant programs discussed by DARPA at the March 2004 DARPAtech Annual Meeting to determine their applicability to solving Navy FORCEnet capability gaps. Also, it is suggested that the Navy should coordinate with the National Coordination Office for Information Technology Research and Development and its recent publication, *Grand Challenges: Science, Engineering, and Societal Advances Requiring Networking and Information Technology Research and Development.*[11]

## 6.5 SUMMARY TECHNOLOGY FINDINGS AND RECOMMENDATIONS FOR THE FORCEₙₑₜ INFORMATION INFRASTRUCTURE

Section 6.2 in this chapter describes science and technology issues associated with achieving FORCEnet. Eight critical FORCEnet functional capabilities are identified, and findings and recommendations are enumerated. Section 6.3.1 addresses the FORCEnet implications of the GIG identifying several key areas that are major S&T issues for the Navy. Section 6.4 addresses the S&T of the ONR program. The committee studied this information and performed a preliminary analysis of findings to determine a recommended priority for the Navy. This analysis included an assessment for each finding of how critical it was for Navy success in executing FORCEnet, whether it was being addressed by others outside the Navy, what potential it had for enhancing performance, and whether it was needed for near-, mid-, or far-term experimentation to assess naval capability.

---

[11] Interagency Working Group on Information Technology Research and Development. 2003. *Grand Challenges: Science, Engineering, and Societal Advances Requiring Networking and Information Technology Research and Development,* Office of Advanced Scientific Computing Research, Department of Energy, Washington, D.C., November (first printing), March 2004 (second printing).

## 6.5.1 Process for the Conduct of
## an Overall Science and Technology Program

In addition to the following prioritized recommendations specific to the needs of the various technology areas described, the ONR will also need to reexamine its process for conducting an overall S&T program that matches the needs of network-centric operations. For ONR to consistently identify S&T gaps, there must be a consolidated set of prioritized FORCEnet capability needs. Regarding process, the committee presents the following findings and recommendations.

### 6.5.1.1 Process Findings

• Today, the processes for identifying needed operational capabilities are multiple, independent, and uncoordinated.

• There is need for a systematic and vigorous process for identifying enabling technologies to satisfy network-centric functional needs as defined and prioritized by NETWARCOM.

• There is a generally recognized need for naval science and technology activities to avoid duplicative effort by maintaining awareness of DOD, academia, and industry/commercial developments in fields relevant to network-centric operations.

• There is a need to conduct all naval technology developments with consideration for the variable reliability of the naval communications environment in which those technologies will eventually have to perform.

### 6.5.1.2 Process Recommendations

Based upon the findings presented above and on the issues described in this chapter, the committee recommends the following:

• **Recommendation** for ONR: Develop a consistent FORCEnet technology roadmap and list of S&T shortfall assessments for guiding naval S&T investment strategy on the basis of a consistent set of FORCEnet capabilities that is recognized across the Navy. It is also recommended that ONR develop a technology roadmap that responds to the FnII capabilities and addresses near-, mid- and far-term capabilities. As the ONR develops this FnII technology roadmap and associated program, it should:

   —Perform sensitivity analyses to evaluate alternatives,

   —Provide cost–benefit analyses,

   —Assess commercial off-the-shelf applicability, and

   —Identify opportunities for leveraging and providing incentives for participation by industry, academia, and other Services.

## 6.5.2  Summary Findings Regarding Technology for the FORCEnet Information Infrastructure

Section 1.3 in Chapter 1 warns that implementing FORCEnet will involve three challenges: an activity of *unprecedented scope*, an *unprecedented need for robustness*, and *significant difficulties in execution* lie ahead. Chapter 4 deals with management and organizational approaches to the scope and execution of FORCEnet, and Chapter 5 addresses engineering approaches. However, the committee finds not only that current technology is inadequate to provide the needed robustness, but also that execution with existing technology may prevent the full realization of FORCEnet's potential.

Network-centric warfare is appealing if the Navy can build and deploy a robust communications fabric and use it to link sensors, weapons, and decision making over great distances. However, network-centric warfare could be a disaster if the fabric is not robust and the naval units do not maintain the capability to operate with uninterrupted, high-capacity connectivity. Enterprise architectures, such as those that the NCES would provide, create a temptation to depend on remote parts of the enterprise for essential services. This temptation needs to be resisted unless the communications fabric is truly robust. Similarly, the Navy should be sure that the fabric is robust before it sends into harm's way ships that have limited organic defensive capabilities and must depend on other units on the network to detect, track, and deal with threats.

In planning naval S&T investments, distinctions should be drawn between challenges on which others are working and those for which the Navy must take responsibility. Four examples of the reasons why naval networks are not robust today are these: (1) blockage of antenna beams by superstructure, (2) dependence on single relays and difficulties in routing through alternative relays, (3) susceptibility to denial-of-service attacks, and (4) the lack of extremely high frequency communications satellite capacity.

The first reason is an example of a problem that naval S&T needs to solve. New technology will be needed for affordable multifunction antennas that can command unobstructed sites atop a ship's superstructure. The second reason is an example of a problem that affects all Services and is receiving attention from all Services, although the Navy may have to plan the acquisition of relay platforms particularly suitable for maritime operations. The third reason is an example of a problem receiving attention from all Services and from the National Security Agency; this problem will require uniform implementation of solutions across the GIG. The fourth reason is a problem for which the Air Force and the National Reconnaissance Office are developing the technology and the Air Force will acquire the satellites, although the Navy may have to program the acquisition of terminals compatible with the new satellites.

In prioritizing naval S&T investments that respond to the robustness challenge, the highest priority should be given to the issues involved in the first problem, which are unique to the Naval Services. However, this priority attention should not be to the exclusion of investments directed at issues involved in the second and third type, which deserve attention from all Services.

In addition to the critical needs for technology to assure robustness, there are needs for technologies that will permit the full benefits of FORCEnet to be realized. Many of them lie in the area of information management, described in detail in Section 6.2. Although few of these needs are challenges solely to the Naval Services, naval participation in the exploration and implementation of the requisite technologies will help assure that they are appropriate for maritime and littoral operations.

Table 6.3 distinguishes different types of S&T challenges: those that appear essential for the Naval Services in order to realize the promise of FORCEnet and those that would further enhance FORCEnet capabilities. The table further distinguishes the essential challenges for which ONR must shoulder the burden and the challenges that are receiving attention from others. It is important to understand that the most critical requirement is to pursue the technologies that will yield a robust information infrastructure.

Finally, in formulating an investment portfolio, consideration should be given to dates when technologies are needed. Although there is an appropriate desire to achieve quick results and apply them to developmental systems quickly, some technology investments may take time to bear fruit and should be scheduled sufficiently early to mature by the time the technology is needed for incorporation into FORCEnet components.

### 6.5.3 Summary Recommendations Regarding Technology for the FORCEnet Information Infrastructure

Based upon the findings presented above and on the issues described in this chapter, the committee recommends the following:

• **Recommendation** for ONR: Give high priority to technology exploration and prototyping to assure continuous connection of naval units to the GIG, giving the highest priority to those naval-unique challenges that others are unlikely to address, including the following:

—Continuing to develop prototypes that demonstrate solutions to the antenna blockage problem shipboard, such as wide-band multibeam arrays, and alternative relays;

—Aggressively seeking technologies that will permit connecting to submarines at speed and depth.

• **Recommendation** for ONR: After assessing the contributions of DARPA and other Services, give high priority to the remaining issues in these areas:

—MANET routing over multiple alternate paths,

—QoS management and monitoring,

TABLE 6.3  Science and Technology (S&T) Findings of the Committee

| FORCEnet Information Infrastructure (FnII) | Navy Priority | | | | |
|---|---|---|---|---|---|
| | Essential to Naval Services; No Others Working | Essential to Naval Services; Others Working | Enhanced-Performance FnII, Desirable But Not Required | Ongoing Office of Naval Research Efforts | Comments |
| *Communications Technology Challenges (see Section 6.2.1.2)* | | | | | |
| Links/antenna | √ | | | Partial | Issue for ships and submarines |
| QoS/monitor | | √ | | Partial | |
| Protocols/STD | √ | | | Partial | |
| Underwater communications | √ | | | Partial | Speed and depth issues, optical communications |
| Satellite relay alternative | | √ | | √ | |
| Optical communications, marine layer | | | √ | √ | |
| Apertures— disadvantaged users | | √ | | Partial | |
| *Information Management Technology Challenges (see Section 6.2.2.2)* | | | | | |
| Ontology | | √ | | Partial | |
| Information services | | √ | | Partial | |
| Sensor resource management | | √ | | | |
| Distributed fusion | √ | | | Partial | Includes underwater issues |
| User-defined visualization | | √ | | | |
| Enterprise monitor | | √ | | | |
| *Situational Awareness Challenges (see Section 6.2.3.2)* | | | | | |
| Contextual reasoning | | √ | | Partial | |
| Knowledge bases | | √ | | Partial | |
| Interactive hypothesis management | | √ | | | |
| Cognitive interface | | | √ | Partial | |

*continues*

TABLE 6.3 Continued

| FORCEnet Information Infrastructure (FnII) | Navy Priority | | | | |
|---|---|---|---|---|---|
| | Essential to Naval Services; No Others Working | Essential to Naval Services; Others Working | Enhanced-Performance FnII, Desirable But Not Required | Ongoing Office of Naval Research Efforts | Comments |
| *Information Assurance Challenges (see Section 6.2.4.2)* | | | | | |
| Metrics/monitoring | | √ | | Partial | |
| Sharing enabler/ balance | | | √ | | |
| Software trustworthiness | | √ | | √ | |
| Power sources | | √ | | √ | |
| *Modeling and Simulation Challenges (see Section 6.2.5.2)* | | | | | |
| Scaling | | √ | | | |
| Systems engineering | | √ | | | |
| "What if" analysis | | | √ | | |
| *Composability Challenges (see Section 6.2.6.2)* | | | | | |
| Mission management | | | √ | | |
| Readiness monitor | | √ | | | |
| Manpower and training | | | √ | | |
| Collaboration | | √ | | Partial | Automation in unreliable communications |
| *Disadvantaged-User Science and Technology Challenges (see Section 6.2.7.2)* | | | | | |
| Human–machine interface | | √ | | Partial | |
| Customized information | | √ | | Partial | |
| Aperture size/ weight | √ | | | Partial | Small ships, underwater platforms |
| Power sources | | √ | | | |
| *Persistent Intelligence, Surveillance, and Reconnaissance Challenges (see Section 6.2.8.2)* | | | | | |
| Automated adaptive sensor control | | √ | | Partial | |
| Manpower reduction | | √ | | Partial | |
| Networked sensors | √ | | | Partial | E.g., sonobuoys and underwater sensors/vehicles |

NOTE: Acronyms are defined in Appendix C.

—Network protection and recovery,

—Information assurance,

—Connectivity to dismounted units with smaller size and weight antenna apertures, and

—Small, networked sensors for wide-area, inexpensive alerting in difficult denied areas.

• **Recommendation** for ONR: Give high priority to information management in the naval context in order to permit full exploitation of network-centric enterprise services, increasing its investments in ontologies of naval operations, information services, distributed Level 2 through Level 4 fusion, and user-defined visualization.

• **Recommendation** for ONR: Invest, as resources permit, in technologies that would further enhance FORCEnet capabilities after due consideration of alternative sources of technology.

# 7

# Implementation Strategy

## 7.1 IMPLEMENTATION STRATEGY OBJECTIVES

The effective and efficient realization of FORCEnet capabilities requires an implementation strategy on the part of the Navy and the Marine Corps. "Implementation" is taken here in the broad sense of ranging from initial conceptual thinking through the establishment of capability in fielded forces. The previous chapters of this report deal with the various aspects of FORCEnet implementation strategy. From that work, the following set of objectives for the strategy can be abstracted:

• *Provide clarity of purpose.* The implementation strategy should make clear the nature of the capabilities to be developed under the FORCEnet initiative. In particular, important points of scope need to be made clear in order to guide implementation properly.

• *Establish an environment and process for the continuous capability evolution and innovation.* As new technical capabilities are explored, new concepts for their application will be discovered and requirements for additional capabilities will be generated. Furthermore, the security environment in which the naval forces operate will continue to change, necessitating new capabilities on the part of the force. Thus, a "closed-form solution" for the development of FORCEnet capabilities cannot be specified in advance—rather, an evolutionary approach is necessary.

• *Apply a forcewide perspective to materiel development.* The nature of network-centric operation is that all components of the force (weapons, sensors, C2 systems, communications, and so on) will relate to and draw upon one another. Thus, these components cannot be designed alone, but only in relation to the other components.

  • *Integrate with joint developments.* All recent combat operations have been joint, and the extent of joint interaction in military operations is only likely to increase. Thus, FORCEnet operational and materiel capabilities must be developed in a joint context.

These objectives are closely related to one another. The environment and process referred to in the second objective enable the overall realization of FORCEnet capabilities. Addressing the need for a forcewide perspective in materiel development, the third objective bores more deeply into one aspect of the process that warrants particular elaboration, and the fourth objective, calling for the integration of FORCEnet capabilities with joint developments, recognizes that the whole process must couple into larger DOD-wide processes. All consideration of the implementation process must be preceded by a clear understanding of what is sought from the process, which is what the first objective, calling for clarity of purpose, requires.

Drawing on the findings and recommendations presented in previous chapters, the sections below elaborate on the objectives, assess how the Naval Services are doing in realizing them, and indicate what more could be done to achieve them.

## 7.2 CLARITY OF PURPOSE

The definition of FORCEnet is as follows:

[FORCEnet is] the operational construct and architectural framework for naval warfare in the information age that integrates warriors, sensors, networks, command and control, platforms, and weapons into a networked, distributed, combat force that is scalable across all levels of conflict from seabed to space and sea to land.[1]

This definition is adequate as a point of departure for the implementation of FORCEnet capabilities, but further elaboration is necessary in order to provide clarity of purpose to all those involved in the implementation. While the definition is quoted quite widely in the Navy and Marine Corps, the committee found little elaboration on two key phrases in it—"operational construct" and "architectural framework." Put simply, the committee took the operational construct to be

---

[1]VADM Richard W. Mayo, USN; and VADM John Nathman, USN. 2003. "Sea Power 21 Series, Part V: FORCEnet: Turning Information into Power," *U.S. Naval Institute Proceedings,* February, p. 42.

the set of concepts of employment for the naval force, and the architectural framework to be the set of architectures describing the structure and relationship of the components of the force. Nevertheless, further elaboration is required to define what constitutes a satisfactory operational construct and architectural framework.

An additional point requiring elaboration relates to the scope of the definition of FORCEnet. In particular, the definition implies that FORCEnet applies to the entire force, not just to the network or information infrastructure as is sometimes implied in writings on FORCEnet. This fact has significant implications for the necessary scope of the concepts of employment and architectures, as is discussed in the sections below. To be clear on this important point: in this report, FnII refers specifically to that infrastructure component of the force, and the word "FORCEnet" alone refers more generally to the force.

As one final point, the discussion here indicates that FORCEnet—as concepts of employment and architectures—is composed of those processes and descriptive items that guide the implementation and realization of network-centric capabilities in the force rather than being the implemented components themselves. The implemented components are referred to using "FORCEnet" as a modifying term—for example, "the FORCEnet Information Infrastructure."

To provide better clarity of purpose, the committee recommends the following:

- **Recommendation** for OPNAV, NETWARCOM, and MCCDC: Articulate better the meaning of the terms "operational construct" and "architectural framework" in the description of FORCEnet and indicate how FORCEnet implementation measures relate to each of these concepts. (Recommendation 1)
- **Recommendation** for OPNAV, NETWARCOM, and MCCDC: Make clear that FORCEnet applies to the entire naval force and not just to its information infrastructure component. In so doing, the organizations should specifically indicate that the concepts of employment and the architectures developed must apply to the operation of the whole force and not just to its information infrastructure component. (Recommendation 2)

The action parties in Recommendations 1 and 2 are so chosen because they are the primary organizations describing the nature of the FORCEnet initiative.

## 7.3 CONTINUOUS CAPABILITY EVOLUTION AND INNOVATION

### 7.3.1 Functional Areas for FORCEnet Capabilities Implementation

Broadly speaking, the overall process for implementing FORCEnet capabilities centers on five functional areas pertaining to (1) operational concepts, (2) requirements, (3) programs and resources, (4) acquisition, and (5) engineering.[2] The responsibilities in these areas, and the Navy offices assigned these responsibilities, are shown in Table 4.1 in Chapter 4.[3]

The process flow through the functional areas could simply be viewed as a linear sequence, from the development of operational concepts through engineering execution. In fact, the situation is highly iterative owing to the coevolution of operational concepts and technology development and the changing nature of the national security environment. This fact is so fundamental that the committee singles it out by recommending the following:

- **Recommendation** for the CNO and the CMC: Promote as a guiding principle that the realization of FORCEnet capabilities will require a process of continuous evolution, involving the close coordination and coupling of the individual departmental functional processes—operational concept and requirements development, program formulation and resource allocation, and acquisition and engineering execution. (Recommendation 3)

The coupling and iteration of individual departmental functional processes are suggested in Figure 7.1. Note from Table 4.1 that within the Navy, one major organization (and its subordinates) has responsibility for each of the "spheres" in the figure—the CFFC, OPNAV, and the ASN(RDA).

The objective of Recommendation 3 is to make the overall implementation process for FORCEnet capabilities as effective as possible. This means achieving maximum effectiveness for the interactions between the functional areas, as suggested by the arrows in the figure, as well as for the individual functional areas (represented within the spheres in Figure 7.1). The next two subsections examine the individual functional areas and the cross-functional interactions, respectively.

### 7.3.2 Assessment of Individual Functional Areas

#### 7.3.2.1 Operational Concept and Requirements Development

The committee makes the following observations about the functional areas of operational concepts and requirements. These observations refer to the Sea

---

[2]Training, an important additional functional area, is not considered here, since it is outside the scope of the study as specified in its terms of reference (see Appendix A). Clearly, significant attention needs to be paid to training for both the operational and the technical aspects of network-centric operations.

[3]The study also presents implementation responsibilities for the Marine Corps (Table 4.2), but the extent of the study's examination did not allow it to develop an assessment of those responsibilities as it did for the Navy.

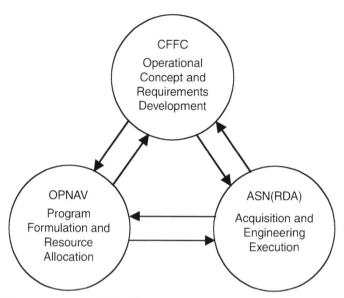

FIGURE 7.1 Implementing FORCEnet.

Power 21 pillars—Sea Basing, Sea Strike, and Sea Shield—as well as to FORCE-
net per se, since FORCEnet capabilities follow from the capabilities required for
the pillars.

• Very little detail has been developed for articulating new operational
concepts—only limited descriptive material and certainly nothing with the sort of
detail typically found in operational architectures.[4] This lack is most likely a
consequence of the very limited resources committed to this area. The Second
and Third Fleets devote only a few people part time to concept development for
the three Sea Power 21 pillars. NETWARCOM appears to have a larger, although
still small, commitment of resources to FORCEnet concept development. Representatives of the said organizations, especially the Second and Third Fleets,
indicated to the committee that these limited commitments were a consequence
of the many demands (e.g., maintaining readiness) placed on these organizations.
• The Second and Third Fleets and NETWARCOM indicated a serious
commitment to experimentation, although generally one of modest scope. The
Second Fleet is active in exploring the use of prototype equipment, the Third
Fleet has a history of experimentation centered around the USS *Coronado* com-

---

[4]In March 2004, after the cutoff date for new input to this study, NETWARCOM initiated an effort
to develop a FORCEnet operational concept, which could provide more detail.

mand ship, and NETWARCOM is conducting the Trident Warrior series of exercises. NETWARCOM's experimentation thus focuses largely on the FnII.

• CFFC has underscored the importance of experimentation by issuing a new experimentation instruction (CFFC Instruction 3900.1A for Sea Trial). Furthermore, it has reduced the number of the large fleet battle experiments to allow more of the smaller, limited objective experiments, which should promote greater exploration and innovation. The Sea Trial instruction promotes greater Navy-wide interaction, thereby potentially bringing more ideas and resources to experiments. At the same time, though, this instruction establishes greater centralized control in approving experiments, which might stifle the very innovation that experimentation seeks to promote.

• NETWARCOM has an active program for FORCEnet requirements development, drawing widespread community participation though its OAG. The OAG does not use any formal analytical methods to relate the requirements to warfighting effectiveness, relying rather on the collective judgment of the group.

• The Second and Third Fleets demonstrated only very limited requirements development for the three Sea Power 21 pillars. This limited work is most likely a consequence of limited resources, as described above for operational concepts development.

Based on the preceding observations, the committee recommends the following:

• **Recommendation** for NETWARCOM, and the Second and Third Fleets especially: Devote significantly more resources to concept development. The criticality of concept development to the overall realization of FORCEnet capabilities certainly requires this increase. The committee also recommends that CFFC determine whether the increased resources would come by reassigning personnel already assigned to the organizations or by request to the CNO for additional personnel. (Recommendation 4)

• **Recommendation** for the CNO: Assign the Pacific Fleet greater direct responsibility in Sea Power 21 concept development. This action would apply the sizable resources and operational experience of Pacific Fleet to help redress the current limitations in resources devoted to concept development. The action would also help strengthen the joint aspects of concept development through Pacific Fleet's relation with PACOM. (Recommendation 5)

• **Recommendation** for the CFFC: Ensure that NETWARCOM plays as broad a role in FORCEnet concept development and experimentation as possible—not just limited to the use of the FnII. This is consistent with NETWARCOM's charter and reflects the fact that FORCEnet refers to forcewide capabilities. (Recommendation 6)

• **Recommendation** for the CFFC: Ensure that the centralized management processes of the new Sea Trial instruction do not stifle innovation. Local initiative is critical to innovation. The Sea Trial management mechanisms should concern themselves with setting broad guidelines and resource allocations within which individual elements in the Navy would be free to innovate. Every experiment, no matter how small, should not require approval by a centralized committee, as would appear to be the case with the new Sea Trial instruction. (Recommendation 7)

• **Recommendation** for NETWARCOM: Develop analytical means for the determination and prioritization of requirements. This would allow requirements to be tied better to warfighting effectiveness and would thereby better support these requirements in the resource-allocation process. (Recommendation 8)

• **Recommendation** for the Second and Third Fleets: Devote more resources to the development of requirements for the three Sea Power 21 pillars. Needed capabilities for the pillars must be adequately specified in order to determine the necessary FORCEnet capabilities. Means to obtain these resources would be addressed by reassigning personnel already assigned to the organizations or by request to the CNO for additional personnel. (Recommendation 9)

### 7.3.2.2 Program Formulation and Resource Allocation

The N6/N7 and N8 use four NCPs—Sea Strike, Sea Shield, Sea Basing, and FORCEnet—in their program formulation and resource-allocation process. Each of these NCPs is further divided into Mission Capability Packages. Those for the FORCEnet NCP are Intelligence, Surveillance, and Reconnaissance; Common Operational and Tactical Pictures; and Networks. As discussed above (Section 7.2), FORCEnet comprises concepts of employment and architectures referring to the entire force, but in the NCP context, the Navy uses the term FORCEnet differently—basically as the FnII (interpreted to include some information-producing assets).

A major shortcoming of the current resource allocation process is thus that it is not configured to implement FORCEnet capabilities, interpreted in their full extent. At most, the process does single out the FnII for resourcing, but in a way that has it competing against the other NCPs, rather than recognizing that the FnII is an enabler of these NCPs.

The problem is further compounded by the limited modeling and simulation tools available to support resource priority decisions. Tools currently used by OPNAV require significant time to set up, thus limiting the number of scenarios that can be examined per budget cycle. In addition, the models and simulations used are of the traditional, attrition-based type, with only limited inclusion of the network and information dimensions of warfare—which are at the heart of FORCEnet capabilities.

Based on these observations, the committee recommends the following:

• **Recommendation** for the N6/N7 and N8: Develop resource-allocation methods directed at realizing forcewide FORCEnet capabilities. Instead of basing the methods on the current Naval Capability Packages, the Navy should instead use "packages" that inherently reflect network-centric operational concepts. FORCEnet Engagement Packs provide one such example. (Recommendation 10)

• **Recommendation** for the N6/N7 and N8: Develop (or acquire) modeling and simulation tools that allow faster exploration of scenarios and better measurement of the effects and limitations of information availability and network connectivity in warfare. This will not be an easy task since such tools are in their infancy, but the Navy should be a proponent for the development of these tools. (Recommendation 11)

### 7.3.2.3 Acquisition and Engineering Execution[5]

All acquisition activity is under the authority of the ASN(RDA). In this capacity, the ASN(RDA) oversees the PEOs, program managers, systems commands, and ONR. In January 2004, the ASN(RDA) led the first meeting of the FORCEnet EXCOMM to address FORCEnet implementation issues. The subjects treated included establishing a FORCEnet implementation baseline and redirecting some current-year funds to support FORCEnet objectives. The decisions and actions of the EXCOMM represent a good start in addressing FORCEnet implementation issues, demonstrating a focus on the future, and communicating an urgency for FORCEnet implementation. The meeting, however, had only very limited attendance from the fleet commands; greater senior-level representation of the fleet command perspective would aid the deliberations at future meetings.

Two significant challenges face the acquisition community in implementing FORCEnet. The first is the need for flexible resource allocation. That is, FORCEnet pertains to the interaction and interoperation of systems across the fleet. Thus, the programs for these systems must be brought forward in as synchronized a manner as possible, which will require the reallocation of funds across programs as problems arise in the development of individual programs and as schedules are adjusted. The EXCOMM's deliberations recognize this problem, but the Navy (as with all of the military Services) is restricted in this regard by the congressional limitations on the transfer of funds between programs.

The second challenge is the need for speed to capability, specifically relating to applying information technologies. That is, the rapid pace of technology change requires that operational users be provided capabilities quickly, before they become obsolete. Furthermore, the whole concept of spiral development implies that system solutions must be developed evolutionarily in concert with operator involvement, and not delivered after the completion of a long-term program without operator involvement. In general, the committee did not get a sense for

---

[5]A significant activity of the engineering component is the development and use of architectures. That subject is treated in Section 7.4.

the rapid development of capability in many of the briefings that it received. Often, no near-term capability delivery (e.g., within 1 year) could be identified. Speed to capability is a matter broader than just the acquisition community (e.g., requirements development is a factor), but greater attention to this matter within the acquisition community is necessary.

The preceding considerations lead the committee to recommend the following:

• **Recommendation** for the ASN(RDA): Take action to include senior members of the fleet commands in the deliberations of the FORCEnet EXCOMM. Their perspective in general would be useful. In particular, the actions necessary to implement FORCEnet capabilities in a fixed-resource environment could impact near-term fleet readiness and should be accomplished in partnership with fleet representatives. (Recommendation 12)

• **Recommendation** for the ASN(RDA): Explore methods for increasing flexibility in resource allocation. One approach for doing so is to aggregate program line items into larger line items, including the possibility of establishing a few major lines referring to FORCEnet capabilities (e.g., for implementation of the FnII or for the systems engineering required across the entire fleet). The Navy, in conjunction with the other military Services, could also consider approaching Congress to relax the limit on reallocating program funds. A strong argument for this authority could be made on the basis of the current need to field systems of systems, in contrast to the previous focus on individual systems. (Recommendation 13)

• **Recommendation** for the ASN(RDA): Review Navy acquisition processes and practices and institute educational measures as necessary, to ensure that programs are providing as rapid a delivery of capability as possible. For example, financial practices could be reviewed to determine means for emphasizing rapid capability delivery while maintaining accountability, and execution instructions could be reviewed to ensure that there is adequate delegation of authority. (Recommendation 14)

### 7.3.3 Assessment of Cross-Functional Interactions

The implementation of FORCEnet capabilities requires an evolutionary approach. Whether one calls this coevolution or spiral development or some similar phrase, what is needed is an effective, highly iterative process coupling the three spheres of responsibility shown in Figure 7.1 above. Such a process requires effective interactions between the individual spheres and among all of them.

#### 7.3.3.1 Discussion of the Interactions

The following describes needed interactions and assesses their realization in current Navy practice.

• *Operational concept and requirements development—program formulation and resource allocation.* In this interaction, the requirements developed under the direction of the CFFC need to be reflected adequately in the program development and prioritization activities of the N6/N7. The FORCEnet requirements developed by NETWARCOM are addressed in the OPNAV programs, but OPNAV uses a priority ranking different from that submitted by NETWARCOM.

• *Program formulation and resource allocation—acquisition and engineering execution.* The programs developed by OPNAV are provided to office of the ASN(RDA) for execution, given budget approval. Beyond that, however, tight interaction between the two organizations is necessary in order to resolve frequent issues of functional allocation and resource adjustments that will arise in program execution, given that FORCEnet relates to the whole fleet. That is, these issues should not be resolved by the acquisition parties alone; they also require the participation of those who developed the rationale and priorities for the programs. Mechanisms to provide this tight coupling between OPNAV and the ASN(RDA) were not apparent to the committee.

• *Acquisition and engineering execution—operational concept and requirements development.* The fleet can provide valuable feedback on the operational utility of assets being acquired by examining them in exercises and experiments. Thus, initial releases or prototypes of systems should be provided to the fleet as early as possible for this purpose. The new Sea Trial instruction from the CFFC appears to be a vehicle to encourage this process.

• *Interaction among all functional areas.* All functional areas must interact effectively to provide speed to capability. Capability needs must be specified, mapped into programs with the appropriate priorities and resourcing, realized through acquisitions, and refined through operational application—all in a timely and flexible manner. As discussed for the acquisition functional area above (Section 7.3.2.3), greater speed to capability is necessary, particularly that pertaining to applying information technologies.

### 7.3.3.2 Means to Improve the Interactions

Making the individual cross-functional interactions and the entire set of interactions work effectively is in good part a matter of governance—giving parties the responsibility and authority to make the interactions and the whole process work. In this regard, the interaction between *program formulation and resource allocation* and *acquisition and engineering execution is considered first.* As discussed above, these two sets of functions need to be more tightly coupled. The committee does not believe that its analyses allow it to make a recommendation for a unique solution to this issue. Rather, it proposes the options presented in the following:

• **Recommendation** for the SECNAV, in conjunction with the CNO and the ASN(RDA): Develop a means to integrate more closely the Navy's program-formulation and acquisition functions, to ensure that adjustments in program execution are consistent with program intent and best serve the overall need of providing forcewide FORCEnet capability. Options to consider include establishing (1) a Programs-Acquisitions Coordination Board co-chaired by the VCNO and the ASN(RDA) or (2) a director of FORCEnet reporting to the VCNO and ASN(RDA). This recommendation envisions that the board or director (depending on which was chosen) would have a major role in carrying out the other recommendations pertaining to program formulation and resource allocation and to acquisition and engineering execution. (Recommendation 15)

The Programs-Acquisitions Coordination Board would have support from a dedicated staff in OPNAV and ASN(RDA). The board would meet regularly, and the staff would work on current issues on a daily basis. The director, FORCEnet, would be a Navy vice admiral[6] and would have a dedicated staff similar to that for a board. The importance and enduring nature of the integration function addressed by Recommendation 15 means the director (or executive secretary of the board) should have a long tenure—6 to 8 years, for example, in contrast to the 2 or 3 years for typical senior assignments.

Recommendation 15 more tightly couples the two lower spheres—*program formulation and resource allocation* and *acquisition and engineering execution*—in Figure 4.1 in Chapter 4. The next matter is the coupling in of the upper sphere, the operational component. In this regard, the committee recommends the following:[7]

• **Recommendation** for the CNO: Charter CFFC to provide periodic assessments of the state of realizing FORCEnet capabilities. The review would include the following: the status and plans for concept development and experimentation for each of the Sea Power 21 pillars and FORCEnet, the current understanding of the set of capabilities required in the fleet, recommended changes in programs to align them better with this set of capabilities, and opportunities for employing acquisition prototypes in naval and joint experiments and exercises. NETWARCOM would provide the staff support to the CFFC in preparing this assessment. (Recommendation 16)

---

[6]Some members of the committee think that the director should be a four-star admiral. This could require the Navy to request an additional four-star billet from the Congress. An argument justifying this billet (and likewise for the other military departments) is that the complexity and importance of system-of-systems issues now facing the departments require an officer of the four-star rank to address them.

[7]Recommendation 12, above, to include senior members of the fleet commands in the deliberations of the ASN(RDA)'s EXCOMM, would also strengthen this coupling.

Beyond these governance measures, there is also a set of "forcing functions" that can be used to drive the implementation process. The intent of these forcing functions is to promote both speed to capability and breadth of capability through a spiral development process. To this end, the committee recommends the following:

- **Recommendation** for the CNO, in conjunction with the ASN(RDA): Establish a set of FORCEnet goals to be realized by specified dates in order to drive the implementation process. Examples of these goals include the provision of specified bandwidth increases and networking capabilities to the fleet, the achievement of designated joint maritime and air situational-awareness capabilities, and the achievement of FORCEnet compliance (or phaseout) for a specified set of legacy systems. Goals could also be of a directly operational nature—for example, the ability to destroy a given class of targets within a stated number of minutes after the targets emerge from hiding. (Recommendation 17)

Further detail and more wide-reaching plans are also necessary to drive the implementation process. In this regard, the committee recommends the following:

- **Recommendation** for the CNO, in conjunction with the ASN(RDA): Direct the preparation of an annual FORCEnet master plan for their review. The plan should lay out milestones—with an emphasis on near-term deliverables—for obtaining key FORCEnet capabilities in terms of operational concepts and systems deployment. The purpose of this plan would be to ensure senior visibility in this area and senior scrutiny of FORCEnet activities and consequent motivation for conducting these activities.[8] (Recommendation 18)

## 7.4 FORCEWIDE PERSPECTIVE FOR MATERIEL DEVELOPMENT

As repeatedly stated, FORCEnet applies to the whole naval force—weapons, sensors, command-and-control systems, communications, and so forth. All of these elements relate to one another in network-centric operations. Thus, the naval force must be designed as a whole—but this does not mean a "grand design" in the sense of traditional systems engineering. The enterprise is far too complex and continual in its development and evolution for that to be possible. Rather, the "lighter touch" of treating the naval force as a complex system applies. This means defining the boundaries ("invariants") that divide the force elements into their major components, broadly partitioning functionality among

---

[8]PEO (C4I & Space) is developing a FORCEnet roadmap that could be used, in part, as the basis for the master plan recommended here.

those components, and establishing interfaces to allow these components to interact with one another. This specification is not permanently fixed, but it will
evolve slowly over time, whereas the technologies used within the components
can change rapidly with time.

The challenge of constructing a complex system requires that the necessary
architectures must be developed. This does not mean that further development of
the naval force should wait several years until a fully detailed architecture is
prepared. Rather, the architectural approach is evolutionary in terms of detail. It
is broadly filled out to start, and then the more detailed pieces are developed as
necessary, and all of these pieces are evolved slowly over time as needed.

Required in conjunction with the architectures are the following: compliance
mechanisms to ensure that the architectures will be applied in program execution,
systems engineering processes to relate the architectures to detailed design and
development, and integration testing to examine whether the system components
actually do interact in the manner intended. The sections below cover this set of
subjects.

### 7.4.1 Architecture Development and Compliance

There are two major architectural efforts pertinent to overall FORCEnet
capabilities—(1) the FORCEnet architecture, developed by SPAWAR,[9] and (2)
the open architecture for combat systems developed by NAVSEA. Volume I of
the SPAWAR architecture is primarily a high-level survey of military systems
that would form components through which the FORCEnet capabilities are realized. Volume II is a list of almost 300 mandated standards applying principally to
the information infrastructure that compliant systems must satisfy. While Volume I generally recognizes the extent of FORCEnet's applicability, neither volume adequately defines the architectural boundaries or invariants, nor provides
mechanisms for allocating functionality among the components. In general, the
committee did not gain an understanding from the architecture documents of how
the pieces described fit together (beyond FnII connectivity) to provide the overall
FORCEnet capabilities.

Working in conjunction with NAVSEA and NAVAIR, SPAWAR is developing a process for assessing compliance of the EXCOMM-directed current systems baseline with the FORCEnet architecture. Generation and assessment of the
baseline represent a sizable effort. Given the limited nature of the current archi-

---

[9]Space and Naval Warfare Systems Command. 2004. *FORCEnet Architecture and Standards,
Volume I, Operational and Systems View*, Version 1.4, San Diego, Calif., April 30; Space and Naval
Warfare Systems Command. 2004. *FORCEnet Architecture and Standards, Volume II, Technical
View*, Version 1.4, San Diego, Calif., April 30. The version of the SPAWAR architecture reviewed
by the committee was Version 1.1, November 18, 2003. A brief review of the later Version 1.4, April
30, 2004, was also made and did not affect the committee's conclusions.

tecture, this assessment will not address matters of overall architectural structure, but it should lead to understanding about the systems changes necessary to enhance information exchange among naval systems.[10]

The NAVSEA open architecture strongly reflects the ideas of invariant boundaries and functional partitioning, although the committee is concerned that the number of boundaries being established could be difficult to maintain.[11] The open architecture represents an important advance in combat systems architecture, and its general design approach is applicable to the FORCEnet architecture as a whole. NAVSEA plans to incorporate the combat systems open architecture into a substantial portion of the fleet by 2008. This effort should establish a modern software architecture that facilitates the maintenance and upgrading of software and allows combat systems to share information more readily with external systems. However, in neither the SPAWAR nor the NAVSEA work reviewed did the committee see a technical discussion of how the FORCEnet and open architectures relate to one another. That understanding is necessary for the combat systems to interface with the external systems.

As a consequence of the preceding observations, the committee recommends the following:

• **Recommendation** for the CNO and the ASN(RDA): Take measures to strengthen the FORCEnet architecture in terms of its ability to represent overall structural relationships among force components. To this end, the CNO and ASN(RDA) should designate NAVSEA, drawing on its open architecture experience, as having a major role in developing the FORCEnet architecture, particularly as pertains to its representation of invariant boundaries and the ability to allocate functionality.[12] Furthermore, SPAWAR and NAVSEA should be directed to specify the technical interrelationship between the FORCEnet architecture and the combat systems open architecture. (Recommendation 19)

The FORCEnet Compliance Checklist,[13] developed under the lead of the N6/N7, lists standards (like those in Volume II of the SPAWAR architecture) that must be obeyed and criteria that must be met in the areas of human–systems

---

[10]The high-level first-phase assessment is planned for completion by September 30, 2004. It is possible that the second-phase assessment planned to follow that will get into matters of functional allocation through the examination of mission threads.

[11]Naval Surface Warfare Center. 2003. *Open Architecture Functional Architecture Definition Document.* Version 2.0, November. For a general discussion of the open architecture initiative, see CAPT Richard T. Rushton, USN, Office of the Chief of Naval Operations (N76); Michael McCrave, ANTEON International Corporation; Mark N. Klett, Klett Consulting Group, Incorporated; and Timothy J. Sorber, Klett Consulting Group, Incorporated, 2004, *Open Architecutre, The Critical Network Centric Warfare Enabler,* Second Edition, Department of the Navy, Washington, D.C., June 29.

[12]Some first-order considerations for defining boundaries are given in Chapter 5.

[13]The version of February 1, 2004, was reviewed by the committee.

integration, spectrum management, information assurance, and joint interoperability. While similar in purpose, a different philosophy is indicated in the Net-Centric Checklist[14] prepared by the ASD(NII). It focuses on capability, whereas the OPNAV checklist focuses on the "what," without reference to the need for a capability. The ASD(NII) checklist is intended to guide and aid the understanding of the design philosophy in system development, whereas the OPNAV checklist gives artifacts that have to be included in the system. The committee prefers the ASD(NII) approach because it should lead to a better understanding of the need for and qualities of the system design. The committee thus recommends the following:

- **Recommendation** for OPNAV: Adopt the Net-Centric Checklist of the ASD(NII) in place of the OPNAV FORCEnet Compliance Checklist, adapting it if necessary to accommodate specific aspects of naval warfare. This design guidance, together with a focus on architectural boundaries, should help promote the development of FORCEnet architectural products. (Recommendation 20)

### 7.4.2 Systems Engineering and Integration Testing

Systems engineering is a process for allocating functionality to subsystems that are bounded by a system architecture, to maximize the effectiveness of the overall system in the set of missions that it is intended to execute. The complexity of the system providing FORCEnet capabilities makes developing and executing a FORCEnet systems engineering process a great challenge. While the principles of traditional systems engineering apply, the FORCEnet process must be expansive in its breadth, with recognition that the complex system will continually develop and evolve.

The systems commands have a long history of systems engineering practice, but in the briefings and other material presented to it, the committee did not see evidence of the major commitment required to meet the FORCEnet systems engineering challenge. SPAWAR is designated as the FORCEnet chief engineer, but the scope of that position is much narrower than the extent of the systems engineering function envisioned by the committee. Thus, the committee recommends the following:

- **Recommendation** for the ASN(RDA), with the support of the systems commands and the relevant PEOs (primarily the PEO C4I & Space and PEO IWS): Develop the capability necessary to effect FORCEnet systems engineering. Very high standards, commensurate with the challenge, should be set for the

---

[14]Office of the Assistant Secretary of Defense for Networks and Information Integration. 2004. *Net-Centric Checklist*, Version 2.1, Department of Defense, Washington, D.C., February 13.

systems engineering staff, who can come from the systems commands, program offices, and outside sources, as necessary. This systems engineering capability would work directly in support of any organization developed to integrate the Navy's program formulation and acquisition functions more closely (as discussed above in Section 7.3.3.2). (Recommendation 21)

On the basis of the experiences of its members in analyzing and developing large-scale systems, the committee recommends the following:

- **Recommendation** for the ASN(RDA), the systems commands, and the relevant PEOs (primarily the PEO C4I & Space and PEO IWS): Pay particular attention to the following in establishing the FORCEnet systems engineering capability:

—Instituting a change management authority responsible for the full set of FORCEnet functional partitions and standards. The decisions of this authority will affect a broad range of naval programs, unprecedented in any prior DOD systems engineering. This authority is key to maintaining the integrity of overall FORCEnet capabilities.

—Providing for the frequent delivery of system capability (e.g., in 6-month increments). This will reinforce the value of the systems engineering process and is in keeping with the need for evolutionary development (discussed above, e.g., Section 7.3.2.3).

—Achieving a mission focus in the analysis and allocation of functionality. For example, each mission can be characterized by a few key variables (e.g., battle force tracking and identification times in antiair warfare) that should be optimized in system design.

—Establishing a rigorous process, independent of individual programs, for recommending the future course of legacy programs—phaseout, retention as is, upgrading, or merger into another program—based on the mission utility of each program.

—Establishing means, involving both process and technology development, to recognize and deal with the vulnerabilities and fragilities that could cause significantly degraded overall capabilities. (Recommendation 22)

The last item in Recommendation 22 represents a particular challenge for complex systems, since their complexity—and in the case of FORCEnet, its unprecedented scope—means that there can be unknown, catastrophic failure modes, much more so than in traditional large-scale systems. Network control, distributed information management, and information assurance are three matters warranting particular attention in this regard for FORCEnet implementation.

Even in the case of a well-architected system with a proper systems engineering process, unanticipated integration difficulties can still occur when the overall system is assembled from its components. Integration testing is thus required to detect and correct such problems. NAVSEA has provided this capability quite effectively for carrier battle groups through the distributed engineering plant. That capability must now be extended to the FORCEnet case. In fact, this capability could also be applied earlier in system development, to provide an exploratory environment for concept development. Thus, the committee recommends the following:

- **Recommendation** for the ASN(RDA), in conjunction with the systems commands and the relevant PEOs (primarily the PEO C4I & Space and PEO IWS): Develop a FORCEnet DEP by generalizing the concepts and approaches used in the current DEP. Since the ongoing joint distributed engineering plant effort does not appear adequate to meet FORCEnet needs (e.g., in terms of scale), the Navy should play a lead role in realizing an extended JDEP by first extending the Navy DEP. (Recommendation 23)

## 7.5 INTEGRATION WITH JOINT DEVELOPMENTS

All recent U.S. combat operations have been joint, and operations will likely only become more so in the foreseeable future. The Navy and Marine Corps are committed to operating as part of a joint force. This viewpoint is made clear in official naval documentation[15] and was expressed by numerous briefers to the committee.

Joint developments refer to the full set of functional areas shown above in Figure 7.1, when taken to apply to all Services and the joint community; the sections below are organized by these three sets of functional areas. Joint developments in these areas are currently in a state of great flux. Policies and processes affecting all areas have recently undergone significant modification, with further refinements being very likely. This flux poses both a challenge and an opportunity for the Naval Services. On the one hand, they will have to keep abreast of the changes and understand their implications for the naval forces. On the other hand, the challenges confronting the DOD as a whole as it seeks to prepare for future environments and institute network-centric capabilities are much the same as those facing the Naval Services, although the DOD's are larger in scope. Thus, a successful FORCEnet implementation strategy can be the model for realizing network-centric capabilities across the DOD.

---

[15]For example, see ADM Vern Clark, USN, Chief of Naval Operations; and Gen Michael W. Hagee, USMC, Commandant of the Marine Corps, 2002, *Naval Operating Concept for Joint Operations*, Washington, D.C., September 22.

## 7.5.1 Operational Concept and Requirements Development

The committee offers the following observations about the functional areas of operational concept and requirements development pertaining to joint operations:

• There is no set of future concepts for joint operations with adequate detail to inform and guide the Navy and Marine Corps in developing their concepts for participating in joint operations. Joint efforts to date, based on JCIDS, have been largely concerned with very broad conceptual development. The Joint Staff recently initiated work on a set of Joint Integrating Concepts that may provide the required specificity.

• The JFCOM joint experimentation process is beginning to transition from being based on JFCOM-originated concepts focused at the operational level of war to becoming a broader process more widely serving the needs of the joint community and Services. Further progress in this direction is necessary, with particular attention on the tactical level of warfare, given the growing joint interaction at that level evident in recent conflicts. In participating in JFCOM experimentation activities, the Navy and Marine Corps need to keep their activities focused so that they do not become overwhelmed by the much greater JFCOM experimentation resources.[16]

• While JFCOM is the executive agent for joint experimentation, the regional combatant commands are becoming an increasing focus for joint concept development and experimentation. The fleet commands are a natural vehicle for interacting with the combatant commands in this regard, as has been the case, such as in the interaction of the Pacific Fleet and its components with the PACOM.

• The derivation of capabilities required by joint forces (and likewise their naval components) is not grounded in the necessary operational and technical analyses. For example, presumed network-centric operational concepts at the tactical level could require network availability and latency that envisioned systems are not technically capable of providing. Furthermore, even if the technical capabilities appear adequate, operational circumstances could lead to degradation in network capabilities. Without the necessary analyses and plans for dealing with the degradations, significant unanticipated shortfalls in deployed capability could well result.

On the basis of the preceding observations, the committee recommends the following:

---

[16]A briefer from NWDC indicated that his organization has 20 individuals assigned to joint experimentation, while JFCOM has 400. Some reallocation of other NWDC billets would appear possible, however.

• **Recommendation** for NETWARCOM, NWDC, and MCCDC: Continue to work with JFCOM to broaden its experimental perspective, with particular emphasis on joint operations at the tactical level. If necessary for these organizations to maintain focused commitment in the face of far larger JFCOM resources, the CFFC, and the Commanding General, MCCDC, should provide guidance on the issues to be addressed and the partitioning of naval involvement in JFCOM, regional combatant command, and Service concept development and experimentation activities. (Recommendation 24)

• **Recommendation** for the fleet commands and MEFs: Build on current interactions with regional combatant commands in order to grow the relationship between naval and joint concept development and experimentation. This means ensuring both that naval concepts are properly embodied in joint concepts and that they reflect the needs of the joint concepts. Combatant command exercises should be used as a principal vehicle for exploring and refining the concepts. This responsibility could require that the fleets devote more resources to concept development (analogous to a point made earlier in Section 7.3.2.1). (Recommendation 25)

• **Recommendation** for NETWARCOM and MCCDC, with technical support from such organizations as SPAWAR and ONR: Undertake a series of naval mission-based analyses to understand the technical limits to achieving network-centric operational concepts and identify approaches for dealing with potential operational degradations in network capabilities. Such analyses should indicate where reliance on more "traditional" capabilities (e.g., the use of localized versus distributed services) may still be necessary, and where increased attention to network path diversity and node heterogeneity is needed to reduce network vulnerability. These results should be shared with other Services and the joint community to increase the understanding of the limits on joint operations.[17] (Recommendation 26)

### 7.5.2 Program Formulation and Resource Allocation

Two topics are considered in this section: (1) matters of broad capabilities definition, prioritization, and resourcing; and (2) the synchronization of naval programs with joint programs providing network-centric capabilities.

### 7.5.2.1 Capabilities Definition, Prioritization, and Resourcing

The traditional defense planning process has been based on requirement statements and corresponding programs developed by the Services. The Joint Defense Capabilities Process instituted by the SECDEF in October 2003 seeks

---

[17]The mission thread analysis planned in conjunction with the FORCEnet baseline assessment could represent a start of the necessary analyses.

instead to establish a process that recognizes from the outset the capability needs for joint warfighting, as determined by the combatant commanders, and provides a means for resourcing those needs. Since the process is in the midst of its first application as of this writing, it is too early to assess what its consequences are. However, if the process is successful in meeting its objectives, there is one clear implication for the Navy and Marine Corps (and other Services): namely, those programs that most contribute to joint warfighting and best serve the needs of the combatant commanders will fare most successfully in the budget progress.

The JCIDS process, established by the CJCS in June 2003, has motivations similar to those of the SECDEF process just noted. The JCIDS process is intended to define and prioritize the capabilities needed for joint operations and to review how well individual programs will contribute to joint operations. While several decisions relating to individual programs have been made, the JCIDS process has, in its year of operation, yet to produce prioritized statements of required capabilities.

Consideration of the SECDEF and CJCS initiatives described above leads the committee to recommend the following:

• **Recommendation** for the N6/N7 and the DCMC(PP&O): Work to articulate clearly how FORCEnet capabilities pertain to joint operations and satisfy the needs of combatant commanders. In the context of the Joint Defense Capabilities Process and JCIDS, this line of argument will strengthen programs providing FORCEnet capabilities in the budget process. While assertions of the joint nature of FORCEnet capabilities have frequently been made by the Navy and Marine Corps in general terms, the committee has not seen any detailed analyses working through the arguments. (Recommendation 27)

• **Recommendation** for the fleet commands and MEFs: Work with the combatant commands to which they are assigned in order to understand and feed into the naval requirements process the capabilities needed by the combatant commands from naval forces. The CFFC, and the Commanding General, MCCDC, would act as the intermediaries for feeding this information from the fleets and MEFs into the program planning processes of the Navy and Marine Corps. (Recommendation 28)

### 7.5.2.2 Synchronization of Naval Programs with Joint Programs

The joint programs under consideration are those that will provide a significantly enhanced GIG over the next several years—e.g., the TSAT system, JTRS, the NCES program, and companion information-assurance programs. The matter confronting the Navy and Marine Corps (and other Services) is one of balance. On the one hand, the GIG programs promise great increases in capability in such areas as communications bandwidth and software services. On the other hand, there are significant risks—for example, TSAT has both technical and budgetary uncertainties, and NCES envisions a services-oriented architecture of unprecedented scope. The Services are thus faced with two issues: (1) how much to rely on capabilities developed under the GIG umbrella in lieu of their own capabilities, and (2) how much to commit now to complementary programs necessary to take advantage of joint GIG programs (e.g., building the ship terminals required for TSAT connectivity, developing NCES-reliant applications).

The Navy has shown a significant awareness of GIG programs and the related issues. The PEO C4I & Space has described benefits, challenges, and options available to the Navy in accommodating GIG capabilities, and ONR has initiated science and technology programs to address naval-unique GIG needs (e.g., ship antennas, maritime battlespace pictures). However, the committee has not seen a strategy articulated at the overall naval program and resource level for relating to joint GIG capabilities. Such a strategy appears warranted, given the capabilities that the joint programs can bring and the substantial naval program costs that could be involved.

On the basis of the discussion above, the committee recommends the following:

• **Recommendation** for the N6/N7 and the Marine Corps Director for C4I: Adopt a prudent course with respect to joint GIG programs, endorsing the further development of these programs but also requiring a clear and continuing assessment of their technical and programmatic progress. In this context, the N6/N7 and the Director, C4I, should clearly understand the limits of applicability of network-centric capabilities, especially at the tactical level (as noted above in Section 7.5.1). (Recommendation 29)

• **Recommendation** for the N6/N7 and N8, and the DCMC(P&R): Articulate programmatic strategies, updated on an annual basis, for leveraging progress and accommodating developments in joint GIG programs. This strategy should lay out approaches for developing the necessary complementary naval capabilities (e.g., terminals, antennas) and describe technical and programmatic alternatives corresponding to the status of joint programs—that is, whether they have remained on schedule, slipped, or failed to meet their objectives. The strategy should also indicate how to leverage joint GIG capabilities as they become available. While some such capabilities will not be deployable for many years (e.g., TSAT), others will be available in the near term (e.g., initial releases of NCES and Horizontal Fusion services). (Recommendation 30)

### 7.5.3 Acquisition and Engineering Execution

The Naval Services should participate actively with joint GIG programs during their execution for two reasons. First, the Naval Services have specific expertise that they can bring to the programs to aid their execution. Current examples of this are SPAWAR's participation in NCES and information assurance developments. Second, the needs of naval (and other Service) missions must be kept constantly in mind as design decisions and trade-offs are made during program execution. TSAT provides a good example of a case in which significant design evolution will likely take place, given the complexity anticipated for the overall system and the fact that the program is in its early stages now. The Navy is selectively involved in joint GIG programs at the present time. To underscore the importance of this involvement, the committee recommends the following:

• **Recommendation** for the ASN(RDA), with support from the PEO C4I & Space, PEO Space Systems, SPAWAR, and MARCORSYSCOM: Track and provide input to the technical development of joint GIG programs to ensure that as these programs evolve, they continue to satisfy naval needs. This objective is best accomplished through naval participation in the programs. The ASN(RDA) should build on current naval participation to ensure that the involvement remains substantive and is across all major GIG programs. The ASN(RDA) should also see that the proper operational perspective (e.g., through the involvement of NETWARCOM) is brought to bear in this activity. (Recommendation 31)

The ONR will need to provide some additional technological solutions even if GIG programs proceed as planned. The GIG will not fulfill some FnII needs, such as communicating with submarines at speed and depth. Furthermore, since NCES as currently envisaged assumes continuous communications connectivity, its use requires some mixture of antenna technology and alternative relay paths to overcome antenna blockages that interrupt connectivity. Fully exploiting enterprise services in a naval context will require considerable exploration and potential technology development. As a consequence of these observations, the committee recommends the following:

• **Recommendation** for ONR: Explore and develop technologies to address naval-unique problems in the GIG context. Areas to be addressed include solving the antenna blockage problem, managing the Navy's inevitably limited bandwidth, connecting submarines at speed and depth to the network, providing alternative communications paths such as high-altitude airborne relays, exploiting the promised enterprise service capabilities of the GIG in the naval context, and constructing consistent maritime battlespace pictures. (Recommendation 32)

A number of policies with related technical documentation (e.g., the GIG architecture) have been instituted, primarily by the ASD(NII), to guide DOD realization of network-centric capabilities. These policies refer to network standards and security, approaches to making data widely available, and use of enterprise software services. In addition, the net-centric review process was initiated early in 2004 by the ASD(NII) to assess the degree to which Service programs are satisfying net-centric compliance criteria and to recommend funding based on the degree of compliance (i.e., reduction or termination in funding for those programs faring poorly against the criteria). These policies and compliance criteria will constrain naval programs, but at the same time they will provide significant external leverage for achieving FORCEnet objectives—that is, realizing network-centric capabilities within the naval forces. The committee thus recommends the following:

• **Recommendation** for the N6/N7, the ASN(RDA), and the MARCORSYSCOM: Fully impose the net-centric criteria mandated by the ASD(NII), in the development and execution of naval programs, subject to any necessary refinement of these criteria. Since the criteria are in their initial use now, the N6/N7, the ASN(RDA), and MARCORSYSCOM should work with the ASD(NII) to refine these criteria as necessary, prior to their full imposition. The use of these criteria will further strengthen related internal policies of the Navy and Marine Corps. Furthermore, if the ASD(NII) net-centric reviews gain strong influence in the DOD budget process, meeting the criteria will be necessary to ensure adequate funding of programs. (Recommendation 33)

The GIG architecture refers primarily to networks and enterprise services, and in operational terms has not been much concerned with the specifics of military missions. Numerous additional architectural developments are underway in the Services, combatant commands, and combat support agencies referring to the more specific concerns of those organizations. The question is how all of these architectures relate to one another. The Navy is involved in limited efforts with other Services to understand such relationships—for example, SPAWAR funding and participation in Air Force and Army architecture development, and NETWARCOM participation in the Joint Rapid Architecture Experimentation initiative promoting horizontal interoperability among the Service's next-generation tactical architectures. While these efforts are useful, the committee believes that more-comprehensive approaches to architectural integration are necessary since, without adequate integration, the overall vision of a network-centric U.S. military force will not be realized. The committee thus recommends the following:

• **Recommendation** for the N6/N7 and the ASN(RDA):[18] Work with OSD and the other Services to develop a better understanding of, and eventually to develop guidelines and principles for, how the numerous architectures being developed in the DOD can be effectively integrated. Particular attention is neces-

---

[18]If a director of FORCEnet were appointed as discussed in Section 7.3.3.2, that individual would be the appropriate party to lead naval efforts in effecting this and the previous recommendation.

sary at the tactical level of warfare, since architectural development of the GIG has not explored that level to a significant extent. The N6/N7 and the ASN(RDA) would be supported by SPAWAR, NETWARCOM, and MARCORSYSCOM in this work. Interaction with the combatant commands (particularly JFCOM and STRATCOM) and combat support agencies (particularly the Defense Information Systems Agency) would also be required. (Recommendation 34)

# 8

# Accounting of the Terms of Reference

The following reproduces in italics the terms of reference of the study with responses and references to specific recommendations in Chapter 7 interpolated.

## TERMS OF REFERENCE

*The Naval Studies Board of the National Academies will conduct a study to assist the Department of the Navy in its approach to implementing FORCEnet. In particular, the study will examine:*

- *The completeness and adequacy of the current definition of FORCEnet, particularly as it supports the concepts presented in Sea Power 21.*

The CNO's definition is complete and adequate, but needs to be more widely understood. See recommendations (1) and (2). FORCEnet is the operational construct and architectural framework for the naval forces. It is more than an information infrastructure.

- *The means for and status of efforts developing and analyzing the operational concepts and associated operational architectures based upon projected FORCEnet capabilities.*

Insufficient resources are being applied to developing and analyzing the operational concept. See recommendations (3) through (8). Program assessments should be based on operational architectures, which usually span more than one of the current Naval Capability Packages. See recommendations (9) and (10).

- *The meaning and mechanisms of "FORCEnet compliance." This aspect of the study will assess and recommend augmenting, as appropriate, current efforts to determine (1) the technical architecture (including standards and associated rules) for FORCEnet, (2) the meaning of compliance for individual force components (e.g., specific platforms, networks), and (3) the mechanisms (e.g., policy directives) for achieving compliance with the architecture.*

FORCEnet compliance is much more than meeting information interoperability specifications. Invariant boundaries must be established along the lines of the NAVSEA open architecture and ASD (NII)'s Net-centric Checklist as the preferable approach to interoperability. See recommendations (18) and (19).

- *The means for achieving the necessary integration and alignment of the various force components—including the technical approaches for the integration, the feasibility of implementing these approaches, the use of modeling and simulation, testing, and the cost implications of so doing.*

ASN(RDA) and PEOs should establish system engineering and integration capabilities (see recommendations (20) and (21)), and, in particular, develop a FORCEnet Distributed Engineering Plant (see recommendation (22)).

- *The adequacy of current organizational responsibilities and relationships in the Navy and Marine Corps for implementing FORCEnet.*

Mechanisms are needed that coordinate activities in three spheres: concept and requirements formulation, resourcing, and acquisition. Fleet representation on the ASN(RDA)'s Executive Committee is needed. Establish either a Board co-chaired by the ASN(RDA) and VCNO or a Director, FORCEnet reporting to CNO and ASN(RDA). See recommendations (11) and (14).

- *The role and types of experimentation needed in developing operational concepts, assessing technical feasibility, and identifying implementation opportunities related to FORCEnet.*

Broaden the role of NETWARCOM beyond that of information infrastructure. See recommendations (5) and (7). Ensure that Sea Trial management processes do not stifle experimentation with emerging technical opportunities. See recommendation (6).

- *The means for leveraging relevant Joint initiatives and those of other Services and the National community, and likewise for influencing those initiatives. Areas considered will include command and control, interoperability, analysis, modeling and simulation, and experimentation.*

Continue to work with JFCOM and regional Combatant Commanders to broaden experimental perspectives and converge naval and joint concept development (see recommendations (23) and (24)) while analyzing the applicability of joint concepts and information infrastructure to naval tactical operations (see recommendations (25) through (28)) and exploiting developments as they become available (see recommendation (29)). Work with the ASD (NII) to understand the capabilities and limitations of the GIG. See recommendation (33).

- *The opportunities for implementing FORCEnet operational concepts and technical capabilities in both the near and longer terms, and means for establishing priorities among those opportunities. Consideration should be given to how to transition such capabilities quickly to the fleet through existing or alternative acquisition mechanisms.*

Seek increased flexibility in resource allocation and streamline internal naval processes. See recommendations (11) through (13). CNO in conjunction with ASN(RDA) should set goals and dates with periodic CFFC progress reviews. See recommendations (15) through (17).

- *The study should consider the above issues in light of ongoing studies and initiatives related to FORCEnet implementation.*

Appendix B lists the studies and initiatives that were formally presented to the committee.

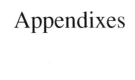

Appendixes

# A

# Biographies of Committee Members and Staff

*Richard J. Ivanetich (Co-Chair)* is an institute fellow at the Institute for Defense Analyses (IDA). His experience spans a number of areas involving defense systems, technology, and operations analyses, having been primarily concerned with computer and information systems, command-and-control systems and procedures, modeling and simulation of systems and forces, crisis management, and strategic and theater nuclear forces. His previous positions at IDA include serving as director of the Computer and Software Engineering Division and assistant director of the System Evaluation Division. Prior to joining IDA in 1975, Dr. Ivanetich was assistant professor of physics at Harvard University. He has served on numerous scientific boards and advisory committees, such as the Defense Advanced Research Projects Agency's (DARPA's) Information Science and Technology Study Group, serving from 1990 to 2004. Dr. Ivanetich was a member of the Naval Studies Board of the National Research Council (NRC) from 1998 to 2004. In 2003 he was elected a National Associate of the National Academies.

*Bruce Wald (Co-Chair)* is founder of Arlington Education Consultants, which advises organizations in both the private and public sectors. Dr. Wald's expertise includes electronic warfare, communications, space surveillance, and computer architectures, with particular emphasis on their implications for naval and national security issues. Dr. Wald served at the Naval Research Laboratory (NRL) for 33 years, in positions of progressively increasing line responsibility in system and technology development, in project and group leadership, and in senior management. In his last position at NRL, Dr. Wald served as associate director of research and director of space and communications technology. Previously, he had been superintendent of NRL's Communications Sciences Division, and before that the founding head of its Computer Science Branch. Dr. Wald has served on panels of numerous scientific boards and advisory committees, such as the Army Science Board, the Defense Science Board, and the Naval Research Advisory Committee. In 2000 he served as chair of the NRC Committee for the Review of the Office of Naval Research's Marine Corps Science and Technology Program. He was a member of the Naval Studies Board from 1994 to 2000, and in 2002 was elected a National Associate of the National Academies.

*Robert F. Brammer* is chief technology officer at TASC, Inc., Northrop Grumman Information Technology, where his responsibilities include leadership and oversight of TASC's technology base and management of the TASC internal research and development program. Dr. Brammer's expertise includes analysis and visualization of very large information security databases and risk-management techniques for distributed critical infrastructure. In addition, he worked on the Apollo program and as a principal investigator on several National Aeronautics and Space Administration satellite remote-sensing programs. Dr. Brammer is a fellow of the Society of Photo-Optical Instrumentation Engineers and of the American Meteorological Society. He has served on numerous scientific boards and advisory committees, such as the Panel on Systems Analysis and Systems Engineering for the National Research Council report *Making the Nation Safer: The Role of Science and Technology in Countering Terrorism* (2002). Dr. Brammer received a Ph.D. in mathematics from the University of Maryland.

*Joseph R. Cipriano* is vice president for advanced solutions at Lockheed Martin Information Technology, where his expertise includes the design, development, and management of large-scale systems and programs. From 1999 until October 2002, Mr. Cipriano served as the Department of the Navy's Program Executive Officer for Information Technology (PEO IT). His efforts in that role led to establishment of the Navy-Marine Corps Intranet (NMCI) Program, the Defense Integrated Military Human Resource System, and the Navy Standard Integrated Personnel System. Prior to serving as PEO IT, Mr. Cipriano served at the Naval Sea Systems Command as the Navy's first Battle Force System Engineer and Deputy Commander for Warfare Systems. In the early 1990s, he was director of the U.S. Department of Energy's Superconducting Super Collider program. Among his many professional awards is the Navy Distinguished Civilian Service Award and the rank of Distinguished Executive in the Senior Executive Service.

*Archie R. Clemins, ADM, USN (Ret.)*, is president of Caribou Technologies, Inc., and co-owner of TableRock International, LLC, both international consulting firms concentrating on the transitioning of commercial technology to government. He retired from the U.S. Navy after more than 30 years of service, concluding that time as commander-in-chief of the U.S. Pacific Fleet, the world's largest combined fleet command. During his Navy service, he strongly supported the establishment of the Navy's Information Technology for the 21st Century (IT-21) and NMCI initiatives. Building on this experience, Admiral Clemins has remained a strong advocate for the accelerated use of information technology and the adaptation of the best commercial practices in the military and the government. Currently, he is vice chair of two start-up firms developing advanced electron beam systems. Admiral Clemins received an M.S. degree in electrical engineering from the University of Illinois, Urbana-Champaign.

*Brig "Chip" Elliott* is principal engineer at BBN Technologies, where he has led the design and successful implementation of a number of secure, mission-critical networks based on novel Internet technology for the United States, Canada, and the United Kingdom. Mr. Elliott's expertise is in the areas of wireless Internet technology, mobile ad hoc networks, quality-of-service issues, and novel routing techniques. He has also acted as a senior adviser on a number of national and commercial networks, including the Discoverer II, Space-Based Infrared System (SBIRS)-Low, and on Celestri/Teledesic satellite constellations and Boeing's Connexion system. He has served on numerous scientific boards and advisory committees, such as the Army Science Board's Advanced Research and Development Activity (ARDA) Technical Experts Panel and the Defense Science Board. Mr. Elliott is a member of the Naval Studies Board.

*Joel S. Engel* is president of JSE Consulting, where he provides guidance to telecommunications equipment companies on next-generation product development and presents expert testimony in legal cases. A member of the National Academy of Engineering, Dr. Engel has expertise in areas including the management of technology strategies, communications system interoperability, network infrastructure development, and cellular phone systems. Dr. Engel's prior positions include those of vice president for technology at Ameritech and vice president for research and development at MCI. A fellow of the Institute of Electrical and Electronics Engineers, Dr. Engel was awarded the Alexander Graham Bell Medal and the National Medal of Technology for his contributions to telecommunications.

*Jude E. Franklin* is technical director for command-and-control systems at Raytheon Network-Centric Systems. His expertise includes strategic technology planning; management of research and development programs; command, control, communications, computers, intelligence, surveillance, and reconnaissance (C4ISR) systems integration; battle management systems; and information technology systems development. Currently, he is responsible for developing strategic technology plans and roadmaps for command-and-control systems. Prior to joining Raytheon, Dr. Franklin spent more than 17 years at Litton PRC, becoming vice president for Litton PRC and chief technology officer and general manager of the PRC Center for Applied Technology. He received a Ph.D. in electrical engineering from Catholic University of America.

*John T. Hanley, Jr.*, is deputy director of the Joint Advanced Warfighting Program at the Institute for Defense Analyses. His expertise includes strategy and concept development, command and control, joint experimentation, military operations analysis, and war gaming. Dr. Hanley formerly served as assistant director for risk management at the Office of Force Transformation of the Secretary of Defense; special assistant to the commander-in-chief, U.S. Forces Pacific; and deputy director of the Chief of Naval Operations Strategic Studies Group. Dr. Hanley holds a Ph.D. in operations research and management science from Yale University.

*Kerrie L. Holley* is chief architect for IBM Global Services e-Business Integration unit, where his expertise includes translating business requirements into process designs for cutting-edge network-centric distributed solutions. An IBM Distinguished Engineer and a member of the IBM Academy of Technology, Mr. Holley has focused on issues related to the modernization of legacy networks and databases to take advantage of Web-services-based computing technologies. Currently, his interests include Web services and e-business solutions, including technical oversight, information technology (IT) consulting, adaptive enterprise architecture design, IT strategy, formation of partnerships among clients and vendors, and management of technical risks. Mr. Holley holds a B.A. degree in mathematics from De Paul University and a J.D. from the De Paul School of Law.

*Kenneth L. Jordan, Jr.*, is an independent consultant. His recent clients include Science Applications International Corporation (SAIC), with which he worked on the assessment of systems engineering and network-centric programs, such as the Wideband Gapfiller Satellite program as well as the Department of Defense Transformational Communications Architecture effort. Previously, Dr. Jordan had spent more than 20 years with SAIC in positions of increasing seniority and oversight, including as corporate vice president and chief scientist of the Information Technology Solutions Group and vice president and division manager of the Command, Control, Communications and Intelligence Systems Analysis Division. In addition, Dr. Jordan served as the principal deputy assistant secretary for research and development, U.S. Air Force, and as director of strategic and space systems in the Office of the Secretary of Defense. Dr. Jordan received an Sc.D. in electrical engineering from the Massachusetts Institute of Technology.

*Otto Kessler* is principal staff engineer at the MITRE Corporation, where his responsibilities include the development of information management systems and technologies focused on the data-gathering and -exploitation needs of the military. Mr. Kessler's expertise includes data fusion, resource and collection management, automated situation and threat analysis techniques, signal and image processing, and development of surveillance sensors and systems. Previously he served at DARPA, the Office of Naval Research, and the Naval Air Development Center. Mr. Kessler's awards include the Defense Medal for Exceptional Public Service.

*Jerry A. Krill* is head of the Power Projection Systems Department at the Johns Hopkins University Applied Physics Laboratory (JHU/APL). The department includes two principal areas: strike warfare (including the JHU/APL work on the Tomahawk Cruise Missile program) and information-centric operations (include JHU/APL's role as trusted agent for systems engineering with the National Security Agency). Dr. Krill's expertise includes weapons systems engineering, sensor-to-weapons networks, missile defense, over-the-horizon missile command-and-control systems, and microwave technology. His prior positions at JHU/APL include programs manager for the Air and Missile Defense Area and supervisor of the Weapon Systems Engineering Branch. Dr. Krill received a Ph.D. in electrical engineering from the University of Maryland.

*Ann K. Miller* is the Cynthia Tang Missouri Distinguished Professor of Computer Engineering at the University of Missouri-Rolla. Her expertise includes information assurance, with an emphasis on computer and network security; and computer engineering, with an emphasis on large-scale systems engineering, satellite communications, and real-time software. Prior to taking up her current position, Dr. Miller had served as Deputy Assistant Secretary of the Navy for Research, Development, and Acquisition (Command, Control, Communications and Intelligence, Electronic Warfare, and Space); Department of the Navy Chief Information Officer; and Director for Information Technologies for Department of Defense Research and Engineering. She received her Ph.D. in mathematics from Saint Louis University.

*William R. Morris, RDML, USN (Ret.)*, retired recently from Pricewaterhouse-Coopers Consulting (PwCC), where he developed and managed business engagements with public- and private-sector clients, including the Assistant Secretary of Defense (Logistics). His expertise includes supply-chain process improvement and planning for both major government organizations and commercial firms. Admiral Morris joined PwCC in 1993 after a distinguished career in the U.S. Navy, during which he held several of the Service's most senior positions in acquisition and logistics, including Competition Advocate General of the Navy and Deputy Assistant Secretary of the Navy (Research, Development, and Acquisition) for business. Admiral Morris is a member of the U.S. Naval Institute.

*Richard J. Nibe, RADM, USN (Ret.)*, retired in May 1999 after 31 years of active duty service in command positions and as a naval aviator. His expertise includes strategic planning and the establishment of policies and priorities for the preparation, execution, and budgeting of large-scale research, development, acquisition, and operation of space-based reconnaissance systems. Admiral Nibe's last assignment involved concurrent appointments as deputy director for military support, National Reconnaissance Office; deputy director for operations (National Systems), J-35, Joint Staff; and deputy director, Defense Support Program Office. Since his retirement, Admiral Nibe has served as an independent consultant on intelligence and defense-related matters. He received a B.S. in aeronautical engineering from the U.S. Naval Academy.

*John E. Rhodes, LtGen, USMC (Ret.)*, retired in August 2000, having served as commanding general of the U.S. Marine Corps Combat Development Command (MCCDC). While at MCCDC, General Rhodes led the Marine Corps in its development of warfighting concepts and in the integration of all aspects of doctrine, organization, training and education, equipment, and support and facilities enabling the Marine Corps to field combat-ready forces. This responsibility entailed, among other things, careful assessments of current and future operating environments and continuous adaptation of the training infrastructure and resources of the Marine Corps in order to ensure that the integrated capabilities were continuously developed for the unified combatant commander.

*Daniel P. Siewiorek* is Buhl University Professor of Electrical and Computer Engineering and Computer Science and director of the Human-Computer Interaction Institute at Carnegie Mellon University. Dr. Siewiorek is a member of the National Academy of Engineering. His expertise includes computer system design automation, methodologies to improve the reliability of computing systems, and wearable, mobile computing systems. At Carnegie Mellon, he leads an interdisciplinary team that has designed and constructed 20 generations of reliable mobile/wearable computing systems. Dr. Siewiorek is a fellow of the Association for Computing Machinery and the Institute of Electrical and Electronics Engineers. He has served on numerous scientific boards and advisory committees such as the NRC's Board on Manufacturing and Engineering Design.

*Edward A. Smith, Jr.*, is senior analyst for effects-based operations and network-centric warfare at the Boeing Company. His expertise includes concept development, naval and defense policy, information warfare, and military intelligence. The career of Dr. Smith, a retired Navy captain with more than 30 years of service, included combat operations in Vietnam and duties as assistant chief of staff for intelligence, Battle Force Sixth Fleet Staff; as deputy director of the Intelligence Directorate, Office of Naval Intelligence; and as intelligence assistant on the Chief of Naval Operations Executive Panel. In addition, Dr. Smith has held positions in the Navy Field Operational Intelligence Office, the Defense Intelligence Agency, and as assistant naval attaché in Paris. He has written broadly on naval operations, and authored a recent book on applying network-centric operations in peace, crisis, and war. Dr. Smith received a Ph.D. in international relations from the American University.

*Michael J. Zyda* is the director of the University of Southern California's (USC's) Viterbi School of Engineering's GamePipe Laboratory, located at the Information Sciences Institute, Marina del Rey, California, and the associate director for games of the USC Integrated Media Systems Center. From fall 2000 to fall 2004, he was the founding director of the MOVES Institute, located at the Naval Postgraduate School (NPS), Monterey, California, and a professor in the Department of Computer Science at NPS as well. From 1986 until the founding of the MOVES Institute, he was the director of the NPSNET Research Group. Professor Zyda's research interests include computer graphics, large-scale, networked 3D virtual environments, agent-based simulation, modeling human and organizational behavior, interactive computer-generated story, modeling and simulation, and interactive games. He is a pioneer in the fields of computer graphics, networked virtual environments, modeling and simulation, and defense/entertainment collaboration. He holds a lifetime appointment as a National Associate of the National Academies, an appointment made by the Council of the National Academy of Sciences in November 2003, awarded in recognition of "extraordinary service" to the National Academies. He served as the principal investigator and development director of the America's Army PC game funded by the Assistant Secretary of the Army for Manpower and Reserve Affairs. He took America's Army from conception to three million plus registered players and hence, transformed Army recruiting.

## Staff

*Charles F. Draper* is director of the NRC's Naval Studies Board. Before joining the NRC in 1997, Dr. Draper was the lead mechanical engineer at S.T. Research Corporation, where he provided technical and program management support for satellite Earth station and small satellite design. He received his Ph.D. in mechanical engineering from Vanderbilt University in 1995; his doctoral research was conducted at the Naval Research Laboratory (NRL), where he used an atomic force microscope to measure the nanomechanical properties of thin-film materials. In parallel with his graduate student duties, Dr. Draper was a mechanical engineer with Geo-Centers, Incorporated, working on-site at NRL on the development of an underwater X-ray backscattering tomography system used for the nondestructive evaluation of U.S. Navy sonar domes on surface ships.

# B

# Agendas for Committee Meetings

**SEPTEMBER 16–17, 2003**
**NATIONAL RESEARCH COUNCIL, WASHINGTON, D.C.**

**Tuesday, September 16, 2003**

**Closed Session: Committee Members and NRC Staff Only**

0830  CONVENE—Welcome, Opening Remarks, Introductions
      —Dr. Richard J. Ivanetich, Committee Co-Chair
      —Dr. Bruce Wald, Committee Co-Chair
      —Dr. Charles F. Draper, Acting Director, Naval Studies Board
      (NSB)
0900  COMPOSITION AND BALANCE DISCUSSION
      —Dr. Dennis Chamot, Associate Executive Director, Division on
      Engineering and Physical Sciences, National Research Council

**Data-Gathering Meeting Not Open to the Public:**
**Classified Discussion (Secret)**

1030  NAVY WARFARE DEVELOPMENT COMMAND—Concepts Development and
      Experimentation Related to FORCEnet
      —Mr. Wayne Perras, Director of Transformation, Navy Warfare
      Development Command

**Closed Session: Committee Members and NRC Staff Only**

1130    COMMITTEE DISCUSSION—Terms of Reference, Study Plans, Other
Issues
—Dr. Richard J. Ivanetich, Committee Co-Chair
—Dr. Bruce Wald, Committee Co-Chair
—Dr. Michael L. Wilson, Program Officer, NSB

**Data-Gathering Meeting Not Open to the Public:**
**Classified Discussion (Secret)**

1300    OFFICE OF THE CHIEF OF NAVAL OPERATIONS: SPACE, INFORMATION WAR-
FARE, COMMAND AND CONTROL DIVISION—Introduction to
FORCEnet and Naval Plans for Integration
—RADM Thomas E. Zelibor, USN, Director, Space, Information
Warfare, Command and Control Division, Office of the Chief
of Naval Operations (OPNAV N61)
1500    MARINE CORPS COMBAT DEVELOPMENT COMMAND—Marine Corps Con-
cepts Development, Experimentation, and Navy Partnering on
FORCEnet
—Mr. James N. Strock, Deputy Director, Expeditionary Force
Development Center, Marine Corps Combat Development
Command
—Mr. Martin Westphal, Director, Command and Control
Integration Division, Expeditionary Force Development Center,
Marine Corps Combat Development Command
—LtCol Timothy J. Jackson, USMC, Deputy Director, Futures
Warfighting Division, Expeditionary Force Development
Center, Marine Corps Combat Development Command
—Mr. James A. Lasswell, Director, Science and Technology
Integration and Future Plans Division, Marine Corps
Warfighting Laboratory

**Closed Session: Committee Members and NRC Staff Only**

1700    COMMITTEE DISCUSSION—Plans Ahead, Report Deliberations
Moderators:
—Dr. Richard J. Ivanetich, Committee Co-Chair
—Dr. Bruce Wald, Committee Co-Chair
1900    END SESSION

**Wednesday, September 17, 2003**

**Closed Session: Committee Members and NRC Staff Only**

0830    CONVENE—Opening Remarks, Committee Discussion, Study Issues, Report Deliberations
          —Dr. Richard J. Ivanetich, Committee Co-Chair
          —Dr. Bruce Wald, Committee Co-Chair
          —Dr. Michael L. Wilson, Program Officer, NSB


**Data-Gathering Meeting Not Open to the Public:**
**Classified Discussion (Secret)**

0900    PROGRAM EXECUTIVE OFFICES—INTEGRATED WARFARE SYSTEMS AND SHIPS—Open Architecture Developments and Other Initiatives
          —Ms. E. Anne Sandel, Executive Director, Program Executive Office for Integrated Warfare Systems
          —Dr. A. Wayne Meeks, Executive Director, Warfare Systems, Naval Sea Systems Command
          —CAPT Thomas Strei, USN, Director, Architecture and Technology, Program Executive Office for Integrated Warfare Systems


**Closed Session: Committee Members and NRC Staff Only**

1000    COMMITTEE DISCUSSION—Meeting Summary, Plans Ahead, Report Deliberations
          Moderators:
          —Dr. Richard J. Ivanetich, Committee Co-Chair
          —Dr. Bruce Wald, Committee Co-Chair


**Data-Gathering Meeting Not Open to the Public:**
**Classified Discussion (Secret)**

1230    ASSISTANT SECRETARY OF DEFENSE FOR NETWORKS AND INFORMATION INTEGRATION—Defense Initiatives Influencing FORCEnet
          —Mr. John L. Osterholz, Director, Architecture and Interoperability, Office of the Deputy Chief Information Officer, Office of the Assistant Secretary of Defense

**Closed Session: Committee Members and NRC Staff Only**

1330     COMMITTEE DISCUSSION—Meeting Summary, Plans Ahead, Report
        Deliberations
          Moderators:
          —Dr. Richard J. Ivanetich, Committee Co-Chair
          —Dr. Bruce Wald, Committee Co-Chair
1530     ADJOURN

**OCTOBER 21–22, 2003**
**NATIONAL RESEARCH COUNCIL, WASHINGTON, D.C.**

**Tuesday, October 21, 2003**

**Closed Session: Committee Members and NRC Staff Only**

0830     CONVENE—Opening Remarks, Committee Discussion
          —Dr. Richard J. Ivanetich, Committee Co-Chair
          —Dr. Bruce Wald, Committee Co-Chair
          —Dr. Michael L. Wilson, Program Officer, NSB

**Data-Gathering Meeting Not Open to the Public:**
**Classified Discussion (Secret)**

0900     SPACE AND NAVAL WARFARE SYSTEMS COMMAND (SPAWAR)—
        FORCEnet Developments and Initiatives
          —Dr. James Kadane, Deputy Chief Engineer, Space and Naval
             Warfare Systems Command
1030     OFFICE OF THE CHIEF OF NAVAL OPERATIONS (OPNAV)—ASSESSMENTS
        DIVISION (N81)—The Role of N81 in the Development of FORCEnet
        Related Programs
          —CDR John C. Oberst, USN, Information Dominance Team
             Lead, Assessments Division, OPNAV N812

**Closed Session: Committee Members and NRC Staff Only**

1230     COMMITTEE DISCUSSION—Topic Area Presentations, Report Delibera-
        tions
          Moderators:
          —Dr. Richard J. Ivanetich, Committee Co-Chair
          —Dr. Bruce Wald, Committee Co-Chair

**Data-Gathering Meeting Not Open to the Public:**
**Classified Discussion (Secret)**

1530    STRATEGIC STUDIES GROUP XXII—The Strategic Studies Group and
        Their Role in the Development of FORCEnet
            —CAPT Joseph N. Giaquinto, USN, Commander, Naval Surface
            Warfare Center, Indian Head Division; Member of Strategic
            Studies Group XXII

**Closed Session: Committee Members and NRC Staff Only**

1630    COMMITTEE DISCUSSION—First Day Summary, Plans Ahead
            Moderators:
            —Dr. Richard J. Ivanetich, Committee Co-Chair
            —Dr. Bruce Wald, Committee Co-Chair
1700    END SESSION

**Wednesday, October 22, 2003**

**Closed Session: Committee Members and NRC Staff Only**

0800    CONVENE—Opening Remarks, Committee Discussion
            —Dr. Richard J. Ivanetich, Committee Co-Chair
            —Dr. Bruce Wald, Committee Co-Chair
            —Dr. Michael L. Wilson, Program Officer, NSB

**Data-Gathering Meeting Not Open to the Public:**
**Classified Discussion (Secret)**

0830    NAVY-MARINE CORPS INTRANET (NMCI)—Impact of FORCEnet on
        Future NMCI Development
            —RADM Charles L. Munns, USN, Director, Navy-Marine Corps
            Internet
1000    NAVAL NETWORK WARFARE COMMAND—Command Overview and
        FORCEnet Developments
            —CAPT John M. Yurchak, USN, Director, Fleet Requirements
            and Assessments, Naval Network Warfare Command
1200    OPNAV—FORCENET WARFARE SPONSOR (N704)—FORCEnet Capa-
        bilities Assessment
            —CAPT Joseph B. Hoeing, USN, Deputy, FORCEnet Warfare
            Integration and Assessments Branch, OPNAV N704B

1330     NAVAL RESEARCH ADVISORY COMMITTEE—NAVY SCIENCE AND TECHNOL-
         OGY IN FORCENET—Study Summary
            —Ms. Teresa Smith, Vice Chair, Committee on Navy Science
              and Technology in FORCEnet, Naval Research Advisory
              Committee

**Closed Session: Committee Members and NRC Staff Only**

1500     COMMITTEE DISCUSSION—Meeting Summary, Plans Ahead
           Moderators:
            —Dr. Richard J. Ivanetich, Committee Co-Chair
            —Dr. Bruce Wald, Committee Co-Chair
1600     ADJOURN

## NOVEMBER 17, 2003
## NAVAL NETWORK WARFARE COMMAND, NORFOLK, VA

### Monday, November 17, 2003

**Closed Session: Committee Members and NRC Staff Only**

0800     CONVENE—Study Update
            —Dr. Richard Ivanetich, Committee Co-Chair
            —Dr. Bruce Wald, Committee Co-Chair
            —Dr. Michael L. Wilson, Program Officer, NSB

**Data-Gathering Meeting Not Open to the Public:**
**Classified Discussion (Secret)**

0830     NAVAL NETWORK WARFARE COMMAND (NETWARCOM)—Welcome,
         Administrative Remarks
            —LCDR Edward Gettins, USN, Office of Requirements,
              NETWARCOM
0835     NETWARCOM—Command Overview; Type Commander Role of
         NETWARCOM; Alignment of NETWARCOM with Respect to Other
         Commands
            —RDML Andrew M. Singer, USN, Deputy Commander,
              NETWARCOM
0915     NETWARCOM—FORCEnet Operational Concepts Development;
         FORCEnet Experimentation Initiatives—Trident Warrior '03
            —CAPT(Sel) Richard Simon, USN, Head, FORCEnet Innovation
              and Experimentation Group, NETWARCOM

1030    NETWARCOM—FORCEnet Requirements and Related Interaction
        with OPNAV
            —CAPT John M. Yurchak, USN, Director, Fleet Requirements
            and Assessments, NETWARCOM
1130    NETWARCOM—Information Operations
            —CDR Steve Carder, USN, Information Operations Architecture,
            Capabilities and Experimentation, NETWARCOM
1200    NETWARCOM—Navy Network Operations
            —Mr. Neal Miller, Deputy for Warfare Requirements,
            NETWARCOM

**Closed Session: Committee Members and NRC Staff Only**

1300    COMMITTEE DISCUSSION—First Day Summary, Group Updates, Report
        Deliberation, Plans Ahead
            Moderators:
            —Dr. Richard J. Ivanetich, Committee Co-Chair
            —Dr. Bruce Wald, Committee Co-Chair
1600    END SESSION


## NOVEMBER 18, 2003
## AIR FORCE COMMAND AND CONTROL, INTELLIGENCE,
## SURVEILLANCE, AND RECONNAISSANCE CENTER;
## COMMANDER, ATLANTIC FLEET COMMAND/COMMANDER,
## FLEET FORCES COMMAND; AND U.S. JOINT FORCES COMMAND,
## NORFOLK, VA

### Tuesday, November 18, 2003

**Data-Gathering Meeting Not Open to the Public:**
**Classified Discussion (Secret)**

0800    AIR FORCE COMMAND AND CONTROL, INTELLIGENCE, SURVEILLANCE, AND
        RECONNAISSANCE CENTER (AF C2ISR) CENTER—Welcome and Intro-
        duction to the Air Force C2ISR Center
            —Maj Gen Tommy F. Crawford, USAF, Commander, Air Force
            C2ISR Center, Langley Air Force Base
0815    AF C2ISR CENTER—Command and Control Constellation;
        CONTROLnet
            —Col Bruce Sturk, USAF, Warfighter Integration, Air Force
            C2ISR Center, Langley Air Force Base
            —Lt Col Richard H. Painter, USAF, Warfighter Integration, Air
            Force C2ISR Center, Langley Air Force Base

0915   AF C2ISR Center—Air Force Experimentation Office—Command and Control Experimentation
    —Mr. Robert Peterman, Air Force Experimentation Office, Air Force C2ISR Center, Langley Air Force Base

1100   Commander, Atlantic fleet/Fleet Forces Command (CLF/CFFC)—CLF/CFFC's Role in the Realization of FORCEnet
    —ADM William J. Fallon, USN, Commander, U.S. Fleet Forces Command; Commander, U.S. Atlantic Fleet

1300   U.S. Joint Forces Command (JFCOM)—Joint Battle Management Command and Control
    —Brig Gen Marc "Buck" Rogers, USAF, Director, Requirements and Assessments, J8 JFCOM

1400   U.S. Joint Forces Command—Collaborative Information Environment (CIE) and Prototyping
    —Mr. Keith Curtis, Collaborative Information Environment Office, JFCOM

1430   U.S. Joint Forces Command—Joint Concept Development and Experimentation
    —CAPT (Sel) Paul Smith, USN, Joint Concept Development and Experimentation Office, JFCOM

1530   U.S. Joint Forces Command—Joint and Naval Transformation Alignment Toward Network-Centric Operations
    —ADM Edmund P. Giambastiani, Jr., USN; Supreme Allied Commander, Transformation; Commander, U.S. Joint Forces Command

1600   U.S. Joint Forces Command—Joint Lessons Learned from Operation Iraqi Freedom
    —Lt Col Mains, USAF, Deputy Director, Joint Lessons Learned Team, U.S. Joint Forces Command

1700   End Session

## NOVEMBER 19, 2003
## U.S. ARMY TRAINING AND
## DOCTRINE COMMAND FUTURES CENTER
## AND U.S. SECOND FLEET, NORFOLK, VA

### Wednesday, November 19, 2003

**Data-Gathering Meeting Not Open to the Public:**
**Classified Discussion (Secret)**

0900   U.S. Army Training and Doctrine Command—Futures Center—Role of the Futures Center for Development of Future Army Warfighting

       —MG Robert W. Mixon, Jr., USA, Deputy Director, Futures
           Center, U.S. Army Training and Doctrine Command
1000    U.S. ARMY TRAINING AND DOCTRINE COMMAND—FUTURES CENTER—The
        Influence of the Futures Center on Army Experimentation, Concepts
        Development, and Networking Initiatives
          —Discussion Participants to Be Determined
1200    U.S. SECOND FLEET—The Role of Second Fleet in Support of Sea
        Power 21, and Sea Strike and Sea Basing Operational Requirements
        Development and Relationship to FORCEnet
          —VADM Gary Roughead, USN, Commander, Second Fleet;
            Commander, Striking Fleet Atlantic; Commander, Joint Task
            Force 120

**Closed Session: Committee Members and NRC Staff Only**

1400    COMMITTEE DISCUSSION—Meeting Summary, Plans Ahead, Report and
        Topic Group Discussions
        Moderators:
          —Dr. Richard J. Ivanetich, Committee Co-Chair
          —Dr. Bruce Wald, Committee Co-Chair
1430    ADJOURN

### DECEMBER 15, 2003
### SPACE AND NAVAL WARFARE SYSTEMS COMMAND,
### AND PROGRAM EXECUTIVE OFFICE FOR
### COMMAND, CONTROL, COMMUNICATIONS, COMPUTERS
### AND INTELLIGENCE, AND SPACE
### SAN DIEGO, CA

**Monday, December 15, 2003**

**Closed Session: Committee Members and NRC Staff Only**

0745    CONVENE—Study Update
          —Dr. Richard J. Ivanetich, Committee Co-Chair
          —Dr. Bruce Wald, Committee Co-Chair
          —Dr. Michael L. Wilson, Program Officer, NSB

**Data-Gathering Meeting Not Open to the Public:**
**Classified Discussion (Secret)**

0800    SPAWAR—Welcome, Command Overview
          —Mr. Scott Randal, Deputy Commander, SPAWAR

0815 SPAWAR—Architecture and Standards; Joint Initiatives; and Current
FORCEnet Assessment Process
—RDML David Antanitus, USN, C4I Chief Engineer, SPAWAR
—CAPT Jim Gosnell, USN, Chief of Operations, C4I Chief
Engineer, SPAWAR
—Dr. Bill Rix, Director, Analysis and Assessments, C4I Chief
Engineer, SPAWAR

1030 SPAWAR—Discussion on the Role of SPAWAR in Developing
FORCEnet
—Mr. Scott Randal, Deputy Commander, SPAWAR
—RDML David Antanitus, USN, C4I Chief Engineer, SPAWAR

1230 PEO C4I & SPACE—Technical Program Development Initiatives in
Support of Naval Operations
—Mr. Andrew Cox, Technical Director, Reusable Application
Integration Development Standards (RAPIDS), PEO
C4I & SPACE
—Mr. Howard Pace, Technical Director, PEO C4I & SPACE

1400 PEO C4I & SPACE—Satellite Communications and Information
Security Issues Influencing FORCEnet
—CDR Steve McPhillips, Deputy Program Manager, Information
Systems Security Program, PEO C4I & SPACE (PMW 161)
—Ms. Michelle E. Bailey, Navy Satellite Communications
Program Manager, PEO C4I & SPACE (PMW 176)

1445 PEO C4I & SPACE—Command Overview; Role of the Program
Executive Officers in Supporting FORCEnet; Alignment of PEO
C4I & SPACE with Respect to Other Commands
—Mr. Dennis Bauman, Program Executive Officer for
C4I & SPACE

**Closed Session: Committee Members and NRC Staff Only**

1600 COMMITTEE DISCUSSION/WORKING DINNER—First Day Summary, Group
Updates, Report Deliberation, Plans Ahead
Moderators:
—Dr. Richard J. Ivanetich, Committee Co-Chair
—Dr. Bruce Wald, Committee Co-Chair

2000 END SESSION

## DECEMBER 16, 2003
## THIRD FLEET; AND 1ST MARINE DIVISION
## SAN DIEGO, CA; AND CAMP PENDLETON, CA

### Tuesday, December 16, 2003

**Data-Gathering Meeting Not Open to the Public:**
**Classified Discussion (Secret)**

0800    THIRD FLEET—Welcome, Command Overview; The Role of Third
        Fleet in Support of Sea Power 21
            —RADM, John R. Hines, Jr., USNR, Deputy Commander, Third
            Fleet
0900    THIRD FLEET—Fleet Experimentation Initiatives Related to FORCEnet
            —CAPT Ellen Jewett, USN, Assistant Chief for Innovation and
            Experimentation, Third Fleet (J9)
            —CAPT Mary Anderson, USN, Assistant Chief for C4I, Third
            Fleet (J6)
            —Col Scott Slater, USMC, Assistant Chief for Plans and Policies,
            Third Fleet (J5)

**Closed Session: Committee Members and NRC Staff Only**

1100    COMMITTEE DISCUSSION/WORKING LUNCH—Group Updates, Report
        Deliberation, Plans Ahead
            Moderators:
            —Dr. Richard J. Ivanetich, Committee Co-Chair
            —Dr. Bruce Wald, Committee Co-Chair

**Data-Gathering Meeting Not Open to the Public:**
**Classified Discussion (Secret)**

1400    1ST MARINE DIVISION—Welcome; Command Overview; Discussion on
        Marine Corps Operations in Iraq and How They Were Influenced by
        Networks and Information
            —Maj Gen James N. Mattis, USMC, Commanding General, 1st
            Marine Division
            —BrigGen John F. Kelly, USMC, Assistant Division
            Commander, 1st Marine Division
1445    1ST MARINE DIVISION—Communications, Information Systems, and
        Operational Lessons Learned
            —LtCol Paul Miller, USMC, Communications (G-6), 1st Marine
            Division

—LtCol Michael Groen, USMC, Intelligence (G-2), 1st Marine Division

—LtCol Clarke Lethin, USMC, Operations and Training (G-3), 1st Marine Division

—LtCol Gary Smythe, USMC, Fire Support Coordinator (G-3 Fires), 1st Marine Division

1700 END SESSION

## DECEMBER 17, 2003
## SPAWAR SYSTEMS CENTER SAN DIEGO, SAN DIEGO, CA

### Wednesday, December 17, 2003

**Data-Gathering Meeting Not Open to the Public: Classified Discussion (Secret)**

0800 SPAWAR SYSTEMS CENTER, SAN DIEGO (SSC-SD)—Welcome, Command Overview; SSC-SD Programs and Initiatives in Support of FORCEnet

—CAPT Timothy V. Flynn, USN, Commander, SPAWAR Systems Center San Diego

—Mr. Rod Smith, Executive Director, SPAWAR Systems Center San Diego

0900 SSC-SD—Composable FORCEnet-Enabled Undersea Warfare Demonstration Facility

—Mr. Jeff Grossman, Deputy for Advanced Technology, Command and Control Department, SSC-SD

—Mr. Michael Reilley, Chief Engineer, Command and Control Department, SSC-SD

1100 ADJOURN

## JANUARY 12–13, 2004
## NATIONAL RESEARCH COUNCIL, WASHINGTON, D.C.

### Monday, January 12, 2004

**Closed Session: Committee Members and NRC Staff Only**

0800 CONVENE—Opening Remarks, Committee Discussion

—Dr. Richard J. Ivanetich, Committee Co-Chair

—Dr. Bruce Wald, Committee Co-Chair

—Dr. Michael L. Wilson, Program Officer, NSB

**Data-Gathering Meeting Not Open to the Public:**
**Classified Discussion (Secret)**

0900 NAVY WARFARE DEVELOPMENT COMMAND—Update on NWDC Concepts Development and Experimentation Related to FORCEnet
 —RADM John M. Kelly, USN, Commander, Navy Warfare Development Command
 —Mr. Wayne Perras, Director of Transformation, Navy Warfare Development Command
1030 ASSISTANT SECRETARY OF THE NAVY FOR RESEARCH, DEVELOPMENT, AND ACQUISITION—Acquisition Initiatives Influencing Fast-Moving Technology Acquisition and FORCEnet
 —Mr. Michael F. Jaggard, Chief of Staff/Policy for the Deputy Assistant Secretary of the Navy for Acquisition Management

**Closed Session: Committee Members and NRC Staff Only**

1200 COMMITTEE DISCUSSION—Report Outline Deliberations
 Moderators:
 —Dr. Richard J. Ivanetich, Committee Co-Chair
 —Dr. Bruce Wald, Committee Co-Chair
1800 END SESSION

**Tuesday, January 13, 2004**

**Closed Session: Committee Members and NRC Staff Only**

0800 CONVENE—Opening Remarks, Report Deliberations
 —Dr. Richard J. Ivanetich, Committee Co-Chair
 —Dr. Bruce Wald, Committee Co-Chair
 —Dr. Michael L. Wilson, Program Officer, NSB

**Data-Gathering Meeting Not Open to the Public:**
**Classified Discussion (Secret)**

1100 NAVAL SEA SYSTEMS COMMAND—Platforms and Systems Integration in Support of Network-Centric Operations and Planning; Status of the Virtual SYSCOM
 —VADM Phillip M. Balisle, USN, Commander, Naval Sea Systems Command

**Closed Session: Committee Members and NRC Staff Only**

1230    COMMITTEE DISCUSSION—Report Deliberations
            Moderators:
            —Dr. Richard J. Ivanetich, Committee Co-Chair
            —Dr. Bruce Wald, Committee Co-Chair

**Data-Gathering Meeting Not Open to the Public: Classified Discussion (Secret)**

1330    JOINT STAFF—COMMAND, CONTROL, COMMUNICATIONS AND COMPUTER
            SYSTEMS—Joint Initiatives Supporting Development of Network-
            Centric Operations
            —LtGen Robert M. Shea, USMC, Director, Command, Control,
               Communications and Computer Systems, The Joint Staff (J6)

**Closed Session: Committee Members and NRC Staff Only**

1430    COMMITTEE DISCUSSION—Report Deliberations, Meeting Summary,
            Plans Ahead
            Moderators:
            —Dr. Richard J. Ivanetich, Committee Co-Chair
            —Dr. Bruce Wald, Committee Co-Chair
1530    ADJOURN

**MARCH 1–5, 2004**
**ARNOLD AND MABEL BECKMAN CENTER, IRVINE, CA**

**Monday, March 1, 2004**

**Closed Session: Committee Members and NRC Staff Only**

0830    CONVENE—Welcome, Administrative Issues, Meeting Schedule
            —Dr. Richard J. Ivanetich, Committee Co-Chair
            —Dr. Bruce Wald, Committee Co-Chair
            —Dr. Michael Wilson, NSB Program Officer
0900    CHAPTER 1 BRIEF—Transforming the Navy/Marine Corps into a Net-
            work-Centric Force
            —Mr. Chip Elliott, Chapter 1 Lead Author
0945    CHAPTER 2 BRIEF—FORCEnet
            —Mr. Ed Smith, Chapter 2 Lead Author
1030    CHAPTER 3 BRIEF—Joint Network-Centric Plans and Initiatives
            —Dr. John Hanley, Chapter 3 Lead Author

### March 2–5, 2004

### Closed Sessions: Committee Members and NRC Staff Only

# C

# Acronyms and Abbreviations

| | |
|---|---|
| ACAT | acquisition category |
| ACO | Allied Command Operations |
| ACSC | Aegis Combat System Center (Wallops Island, Maryland) |
| ACT | Allied Command Transformation |
| ADNS | Advanced Digital Network System |
| AF C2ISR | Air Force Command and Control, Intelligence, Surveillance, and Reconnaissance |
| AODV | Ad hoc On-demand Distance Vector |
| ASD(NII) | Assistant Secretary of Defense for Networks and Information Integration |
| ASN(RDA) | Assistant Secretary of the Navy for Research, Development, and Acquisition |
| ASW | antisubmarine warfare |
| ATO | Air Tasking Order |
| | |
| BFT | Blue Force Tracking |
| | |
| C2 | command and control |
| C4I | command, control, communications, computers, and intelligence |
| C4ISR | command, control, communications, computers, intelligence, surveillance, and reconnaissance |
| C5I | command, control, communications, computers, combat direction, and intelligence |

| | |
|---|---|
| CD&E | concept development and experimentation |
| CERT/CC | Computer Emergency Response Team Coordination Center |
| CES | Core Enterprise Services |
| CFFC | Commander, Fleet Forces Command |
| CINC | Commander-in-Chief |
| CJCS | Chairman of the Joint Chiefs of Staff |
| CMC | Commandant of the Marine Corps |
| CNO | Chief of Naval Operations |
| CNR | Chief of Naval Research |
| COI | community of interest |
| COP | common operational picture |
| COTP | common operational and tactical picture |
| COTS | commercial off-the-shelf |
| CVN | aircraft carrier |
| | |
| DARPA | Defense Advanced Research Projects Agency |
| DCGS | Distributed Common Ground/Surface Systems |
| DCMC(I&L) | Deputy Commandant of the Marine Corps for Installations and Logistics |
| DCMC(P&R) | Deputy Commandant of the Marine Corps for Programs and Resources |
| DCMC(PP&O) | Deputy Commandant of the Marine Corps for Plans, Policies, and Operations |
| DCNO | Deputy Chief of Naval Operations |
| DEP | distributed engineering plant |
| D&I | Discovery and Invention |
| DISA | Defense Information Systems Agency |
| DOD | Department of Defense |
| DON | Department of the Navy |
| DOTMLPF | doctrine, organization, training, materiel, leadership and education, personnel and facilities |
| DSB | Defense Science Board |
| DTN | Delay-Tolerant Networking |
| | |
| EMI | electromagnetic interference |
| EMW | Expeditionary Maneuver Warfare |
| EPP | Enhanced Planning Process |
| ESF | Expeditionary Strike Force |
| ESG | Expeditionary Strike Group |
| ETA | estimated time of arrival |
| EXCOMM | Executive Committee |

| | |
|---|---|
| FBE | fleet battle experiment |
| FCS | Future Combat Systems |
| FNC | Future Naval Capabilities |
| FnII | FORCEnet Information Infrastructure |
| | |
| GCCS | Global Command and Control System |
| GIG | Global Information Grid |
| GIG-BE | Global Information Grid-Bandwidth Expansion |
| GIG ES | Global Information Grid Enterprise Services |
| GMTI | ground moving target indication |
| GRA | Government Reference Architecture |
| | |
| HAIPE | High Assurance Internet Protocol Encryptor |
| HLA | high-level architecture |
| | |
| INMARSAT | International Marine/Maritime Satellite |
| INTEL | intelligence |
| IP | Internet Protocol |
| IPD | integrated product demonstration |
| IPv6 | Internet Protocol, Version 6 |
| ISR | intelligence, surveillance, and reconnaissance |
| IT | information technology |
| IT-21 | Information Technology for the 21st Century (Navy program) |
| IV&V | independent verification and validation |
| IWS | Integrated Warfare Systems |
| | |
| JBMC2 | joint battle management command and control |
| JCD&E | Joint Concept Development and Experimentation |
| JCIDS | Joint Capabilities Integration and Development System |
| JDAM | Joint Direct Attack Munition |
| JDEP | Joint Distributed Engineering Plant |
| JFC | Joint Force Commander |
| JFCOM | U.S. Joint Forces Command |
| JFMCC | Joint Force Maritime Component Commander |
| JFN | Joint Fires Network |
| JIWP | Joint Integrated Warfare Picture |
| JNTC | joint national training capability |
| JOC | joint operating concept |
| JOpsC | Joint Operations Concept |
| JROC | Joint Requirements Oversight Council |
| JSF | Joint Strike Fighter |
| JTRS | Joint Tactical Radio System |

| | |
|---|---|
| LCS | Littoral Combat Ship |
| LOE | limited objective experiment |
| | |
| MAGTF | Marine Air-Ground Task Force |
| MANET | Mobile Ad-Hoc Network |
| MARCORSYSCOM | Marine Corps Systems Command |
| MBC | Maritime Battle Center |
| MCCDC | Marine Corps Combat Development Command |
| MCEN | Marine Corps Enterprise Network |
| MCM | mine countermeasures |
| MCP | Mission Capability Package |
| MCWL | Marine Corps Warfighting Laboratory |
| MEF | Marine Expeditionary Force |
| MIW | mine warfare |
| MLS | multilevel security |
| MN | multinational |
| M&S | modeling and simulation |
| | |
| N6/N7 | Deputy Chief of Naval Operations for Warfare Requirements and Programs |
| N8 | Deputy Chief of Naval Operations for Resources, Requirements, and Assessments |
| NATO | North Atlantic Treaty Organization |
| NAVAIR | Naval Air Systems Command |
| NAVSEA | Naval Sea Systems Command |
| NCDP | Naval Capabilities Development Process |
| NCES | Network-Centric Enterprise Services |
| NCO | network-centric operations |
| NCOW RM | Network-Centric Operations and Warfare Reference Model |
| NCP | Naval Capability Pillar |
| NCTAMS | Naval Computer and Telecommunications Area Master Station |
| NETWARCOM | Naval Network Warfare Command |
| NGA | National Geospatial-intelligence Agency |
| NII | Networks and Information Integration |
| NMCI | Navy-Marine Corps Intranet |
| NNEC | NATO Network-Enabled Capability |
| NNP | Naval Nuclear Propulsion |
| NORTHCOM | U.S. Northern Command (Homeland Security) |
| NRC | National Research Council |
| NRO | National Reconnaissance Office |

| NTDS | Naval Tactical Data System |
|------|---------------------------|
| NWDC | Navy Warfare Development Command |
| | |
| OA | open architecture |
| OACE | Open Architecture Computing Environment |
| OAG | Operational Advisory Group |
| OHIO | only handle information once |
| OIF | Operation Iraqi Freedom |
| OLSR | Optimized Link State Routing |
| ONR | Office of Naval Research |
| OPNAV | Office of the Chief of Naval Operations |
| OSD | Office of the Secretary of Defense |
| | |
| PACOM | Pacific Command |
| PDA | personal digital assistant |
| PEO | program executive office/officer |
| POM | Program Objective Memorandum |
| POSIX | Portable Operating System for Information Exchange |
| | |
| QoS | quality of service |
| | |
| R&D | research and development |
| RF | radio frequency |
| RFP | request for proposal |
| RTO | Research and Technology Organization |
| | |
| SBR | space-based radar |
| SEAL | Sea, Air, Land (U.S. Navy military special forces team member) |
| SECDEF | Secretary of Defense |
| SECNAV | Secretary of the Navy |
| SJFHQ | Standing Joint Force Headquarters |
| SOA | Services Oriented Architecture |
| SOCOM | Special Operations Command |
| SPAWAR | Space and Naval Warfare Systems Command |
| SPG | Strategic Planning Guidance |
| SSG | Strategic Studies Group |
| S&T | science and technology |
| STIMS | Sea Trial Information Management System |
| STRATCOM | U.S. Strategic Command |
| SUW | surface warfare |
| SYSCOM | Systems Command |

| | |
|---|---|
| TAD | Theater Air Defense |
| TCA | Transformational Communications Architecture |
| TCS | Transformational Communication System |
| TMD | Theater Missile Defense |
| TOP | tactical operational picture |
| TPPU | task, post, process, use |
| TSAT | Transformational Satellite |
| TTPs | tactics, techniques, and procedures |
| | |
| UAV | unmanned air vehicle |
| UHF | ultrahigh frequency (300 to 3,000 MHz; 1 m to 10 cm) |
| USMC | U.S. Marine Corps |
| USN | U.S. Navy |
| | |
| VCNO | Vice Chief of Naval Operations |
| | |
| XML | Extensible Markup Language |
| XTCF | eXtensible Tactical C4I Framework |